Advances in Fusion Energy Research - From Theory to Models, Algorithms, and Applications

Edited by Bruno Carpentieri and Aamir Shahzad

Published in London, United Kingdom

Advances in Fusion Energy Research - From Theory to Models, Algorithms, and Applications
http://dx.doi.org/10.5772/intechopen.100766
Edited by Bruno Carpentieri and Aamir Shahzad

Contributors
Javier Resta López, Colin Windsor, Gurdeep Singh Kamal, Jian Rong, Guillaume Urbanczyk, Bruno Carpentieri, Salvatore Ventre, Gaspare Giovinco, Fabio Villone, Guglielmo Rubinacci, Antonello Tamburrino, Christopher Tully, Andi Tan, Wonyong Chung, Mingchao Cai, Mo Mu, Lian Zhang, Aamir Shahzad, Zamar Ahmed, Muhammad Kashif, Amjad Sohail, Alina Manzoor, Fazeelat Hanif, Rabia Waris, Sirag Ahmed

First published in London, United Kingdom, 2022 by IntechOpen
IntechOpen is the global imprint of INTECHOPEN LIMITED, registered in England and Wales, registration number: 11086078, 5 Princes Gate Court, London, SW7 2QJ, United Kingdom

British Library Cataloguing-in-Publication Data
A catalogue record for this book is available from the British Library

Additional hard and PDF copies can be obtained from orders@intechopen.com

Advances in Fusion Energy Research - From Theory to Models, Algorithms, and Applications
Edited by Bruno Carpentieri and Aamir Shahzad
p. cm.
Print ISBN 978-1-80356-278-0
Online ISBN 978-1-80356-279-7
eBook (PDF) ISBN 978-1-80356-280-3

Meet the editors

Bruno Carpentieri obtained a Laurea degree in Applied Mathematics in 1997 from Bari University, Italy. He furthered his Ph.D. studies in Computer Science at the Institut National Polytechnique de Toulouse (INPT), France. He served as an assistant professor at the Bernoulli Institute for Mathematics, Computer Science and Artificial Intelligence, University of Groningen, Netherlands, and as a reader in applied mathematics at Nottingham Trent University, UK. Since May 2017, he has been an Assistant Professor of Applied Mathematics at the Faculty of Computer Science, Free University of Bozen-Bolzano, Italy. His research interests include applied mathematics, numerical linear algebra, and high-performance computing. Dr. Carpentieri has been a member of several scientific advisory boards in computational mathematics. He is an editorial board member of the *Journal of Applied Mathematics*, an editorial committee member of *Mathematical Reviews*, and a reviewer of thirty scientific journals in numerical analysis. He has co-authored around fifty publications in peer-reviewed scientific journals.

Aamir Shahzad has more than seventeen years of experience in university research and teaching at home and abroad. He studied the thermophysical properties of materials and plasmas at graduate and postgraduate levels. He received his post-doctorate and doctoral degrees from Xi'an Jiaotong University (XJTU), China, in 2015 and 2012, respectively. He has proposed novel methods to explore outcomes of complex materials, highlighting his aptitude in computational physics and molecular modeling and simulations. Moreover, Dr. Shahzad is interested in computational physics, complex fluids/plasmas, plasma oncology, and bio- and energy materials. Currently, Dr. Shahzad is an associate professor in the Department of Physics, at GC University Faisalabad (GCUF). He is a member of the XJTU ThermoPhysical Society and the GCUF and the University of Agriculture Faisalabad Physics Societies.

Contents

Preface

This book discusses some recent advances in fusion energy research, including the development of new theory, models, and algorithms, and their application in solving practical computational science and engineering problems. Although it is now established that fusion power can make an enduring contribution to future energy supply, offering many benefits of both renewable and resource-limited energy sources, several challenges still need to be addressed before fusion energy can become a commercial reality. Conducting materials need to have extremely high heat tolerances and low enough vapor pressure to reduce plasma contamination. The extreme temperatures and strong magnetic fields inside a nuclear fusion reactor can make the power generation process highly prone to instabilities. Sudden changes in plasma positions during plasma instabilities may lead to disruptions characterized by a loss of magnetic confinement with subsequent release of the magnetic and thermal energy stored in the plasma to surrounding structures. The development of new models and computational techniques that enable us to control these instabilities is critical to establishing the future of nuclear fusion as an efficient alternative energy source.

From a computational viewpoint, fusion devices are challenging to model. The problem is inherently multiphysics due to the electromagnetic interaction of fusion plasma with the surrounding conducting structures. Although most fusion devices have a nominal toroidal geometry, the geometry of the conductors can be rather complex due to several deviations from this ideal situation (holes, cuts, slits, ports, etc.), for example, to make the interior of the machine accessible. A detailed three-dimensional description of the structures is needed to provide an accurate estimate of the effects of current flows, giving rise to models that can be extremely demanding in terms of the computational burden. Fast parallel techniques are often required to make the computations more affordable.

This book includes contributions on several interesting topics, describing new technology and solutions, presenting the development of new models and algorithms, and discussing both engineering challenges and their respective underlying physics, together with examples of both simulations and experimental results from realistic tokamaks configurations. Researchers, engineers, and graduate students in both pure and applied physics, mathematics, and engineering may benefit from this volume. We express appreciation to IntechOpen for their professional support, especially Author Service Manager Ms. Marica Novaković for her patience and continued help in the preparation of this volume.

Bruno Carpentieri
Faculty of Computer Science,
Free University of Bozen-Bolzano,
Bolzano, Italy

Dr. Aamir Shahzad
Professor (Associate),
Modeling and Simulation Laboratory,
Department of Physics,
Government College University,
Faisalabad, Pakistan

Chapter 1

Introductory Chapter: Large Eddy Simulation for Turbulence Modeling

Aamir Shahzad, Muhammad Kashif and Fazeelat Hanif

1. Introduction

Laminar and non-laminar are two major types of flows and are discussed mainly in fluid mechanics. Streamline and turbulent flows are examples of laminar and non-laminar flows; however, a laminar flow can be transformed into non-laminar by applying some kind of perturbations (such as temperature, pressure, force field, etc.) and/or by employing some gradient. This type of conversion process from laminar to non-laminar flow produces new patterns in between the two states, and these patterns are unstable, and it generates the flow instabilities in the fluid. Only large eddies (large-scale motions) are directly computed in large eddy simulation (LES); therefore, to create the three-dimensional (3D) unsteady governing equations for large-scale motions, a low-pass spatial filter is used for the instantaneous conservation equations. LES has wide range of applications for compressible flows, turbulent combustion, aeroacoustics, turbulent/transitional flows, and atmospheric sciences. In comparison to LES for cases involving incompressible flows, much less work has been performed on LES for compressible flows, and there are numerous difficulties/problems in this field. Extra work/requirements are needed for supersonic flows with shock waves in order to accurately and steadily capture the shock while also providing the spatial accuracy necessary to simulate a number of fine-scale turbulence structures. Low-order techniques are typically used to address shock waves, frequently using upwind schemes that are not particularly suitable for LES. Favre filtering is typically used in compressible flows to prevent the entry of sub-grid scale (SGS) terms into the continuity equation; therefore, knowledge and expertise obtained in incompressible flows may not be applicable. SGS modeling for compressible flows is significantly more difficult as a result of the additional equations that must be solved, such as the energy equation for the compressible case, and the necessity to represent additional SGS terms, such as the SGS heat flux [1].

With applications in a variety of combustion issues, LES of turbulent combustion first emerged in the 1990s and has grown significantly in the last 10 years. The majority of combustion chemistry takes place in SGS; therefore, models must be created because chemical reactions typically take place on sizes much smaller than those of LES meshes. However, even with very straightforward SGS combustion models, LES has showed considerable potential in this field and clearly outperformed the Reynolds-averaged Navier-Stokes (RANS) approach. However, due to the complexity of turbulent combustion, which includes chemical reactions, turbulence/chemistry interactions, liquid fuel atomization, liquid fuel injection, droplet breakup and evaporation, small-scale

IntechOpen

molecular fuel air mixing, large-scale turbulent fuel air mixing, and chemical reactions in aircraft engines, there are significant challenges in this area. Many of these processes take place at various ranges of length and time [1]. LES also has wide range of applications in gas turbine. Reduced life cycle costs for the operator and decreased environmental effect during engine production and operation are crucial factors for gas turbine manufacturers. Many gas turbine components use a stable RANS approach; however, when it comes to the rotor-stator interaction in turbomachinery, we frequently use an unsteady RANS (URANS) approach where the governing flow equations are phase averaged assuming a constant rate of rotation [2]. LES-related techniques are being used more frequently as a result of the inability of RANS models to produce accurate off-design aerodynamics forecasts, noise source information, and predictive capability for the control of drag, noise, and mixing processes in general. The huge advances in accessible computing power over the past few years have provided additional encouragement for the usage of LES [3].

1.1 LES for aeroacoustics

A major amount of the noise that is produced by air and land transportation, such as fan noise, jet noise, high-speed train noise, and airframe noise, is a growing environmental concern. Turbulence is a prominent source of aerodynamic noise. Because LES directly computes large scale fluctuations, which are known to add the most to the noise generated in many issues, LES is a very helpful technique in aeroacoustics. Applications of LES for foretelling aerodynamic noise likely began in the 1990s and have since grown to be a very active area of research. LES shows considerable promise for aeroacoustics computations, from improving source modeling for acoustic analogies to practical prediction and designing of engineering systems in the near future. It also advances fundamental understanding of noise creation. LES algorithms should be able to reliably mimic the flow physics that captures the transfer of energy from turbulent to acoustic modes if properly built and validated. The proper SGS modeling, numerical issues such as high-order accuracy and careful usage of the boundary conditions, and practical engineering configurations where flow Reynolds numbers are typically very high make it impractical to use LES for both noise source capturing and its propagation and are all still significant challenges. Additionally, standard validation study against approved experimental databases can be carried out for relatively basic LES applications. Since intricate statistics such as two-point space-time correlations are essential to flow-generated sound, more attention should be given while validating LES applications in aeroacoustics, according to the theory. As a result, the validation may begin with the most basic facts before moving on to more intricate and acoustically important statistics [1].

1.2 LES for turbulent/transitional flows

Turbulence has an important property, which is scale invariance. Certain characteristics of the flow are said to be scale invariant if they hold true across various motion scales. Such symmetry can be explained as an especially straightforward relation between small and large scales, making it a crucial component in turbulence models [4]. LES, a different strategy, was first suggested by Smagorinsky in 1963. The traditional RANS approach, which requires solving additional modeled transport equations to determine the so-called Reynolds stresses as a result of the averaging procedure, is not used by LES. When compared with direct numerical simulation (DNS), the computational cost of LES is significantly lower since only small-scale SGS motions are represented and the large-scale motions of turbulent flow are computed directly. LES is more perfect than the RANS approach as the large eddies contain most of the turbulent energy and are accountable for most of the turbulent mixing and momentum transfer, and these eddies are captured directly by LES in full detail where they are modeled in the RANS approach. Moreover, the small scales tend to be more homogeneous and isotropic than the large ones, and thus, the SGS motions modeling should be easier than all scales modeling within a single model such as in the RANS approach [1]. There are two main types of SGS models for LES of turbulent flows. The models that yield expressions for SGS terminology such as a heat flux or stress tensor and typically include eddy viscosity notions fall under one group. The SGS stresses are secondary values, which are directly computed from the definitions in the other category, which describes the unresolved primitive variables such as velocity or temperature [5]. For most meteorological applications, large eddies, which hold the majority of the turbulent kinetic energy (TKE) also referred as energy-containing eddies, are the most significant scales for atmospheric planetary boundary layer (PBL) turbulence, as an example, and are in charge of most turbulent transport concerned. It is a simulation that specifically determines (or eliminates) large eddies when LES roughly reflects the effects of smaller ones. As LES's grid resolution increases, less are present, a greater spectrum of turbulent eddies is resolved, and LES-generated flows are parameterized, and they become more representative throughout the flow field. Thus, now LES is the most promising/feasible numerical mean for realistic turbulent/transitional flows simulation [6]. LES inlet conditions were also generated on the basis of implementation of digital filter generator (DFG). The test case was the LES of a channel flow with a continuously repeated constriction. The DFG was used to construct three-time series that would be used as LES inlet conditions in the future. Along with entering the first and second moments of the velocity field over the inlet plane from the periodic boundary condition (PBC) simulation in the first, the DFG's input turbulence scales were set to be spatially homogeneous with values determined using a channel inlet height-weighted area average. The turbulence scales were allowed to change in the second and third time series. Their variation is again inferred from the PBC simulation and is related to wall normal direction. Then, LES's inlet boundary conditions were created using these distinct time series. These changing turbulent scales have increased the simulation accuracy, which has significant uses [7].

1.3 LES for atmospheric science

Prior LES research mostly concentrated on a turbulent PBL with flat terrain and no clouds. Due to the enormous thermal plumes present and the lack of other flow conditions, this flow regime is best suited to LES without involving complex physical phenomena (such as latent heating and radiation). However, in recent years, LES has been broadened to study more challenging and complicated PBL regimes, which are pertinent to forecasts for severe weather and climate, or more recent applications for wind energy [6].

2. History, current state, and future challenges of LES

2.1 History

Smagorinsky first proposed LES in 1963 for forecasting air flow, and early uses likewise fell under this category. Deardoff and Schumann introduced LES to engineering-related flow for the first time in 1970 and 1975, respectively. From the 1960s to roughly the middle of the 1980s, LES took a while to develop, and the applications primarily consisted of straightforward, building-block flows, such as homogeneous turbulence, plane channel flows, mixing layers, and so on. However, as computing power increased, a very fast development and sharp increase in LES applications began around the middle of the 1980s, particularly after the 1990s with substantial growth of the LES community and a wide range of LES applications shifting from simple flows to complex flows, including multi-phase flow, heat transfer, and fluid dynamics, aeroacoustics, transmission, combustion, etc. In addition to the rise in computing power, it is now evident that RANS approaches naturally have limitations and can't deal with certain categories of sophisticated turbulent flow issues, and LES has developed quickly and has a wide range of applications [1].

2.2 Current state

As was indicated in the preceding section, the LES was initially applied successfully to examine the specifics of flow problems with low Reynolds numbers and generally simple geometry, such as homogeneous turbulence, mixing layers, and flat channel flows. Despite the fact that LES is used in such academic or essential setups still exists today, primarily for model validation and a basic comprehension of flow physics, etc. When the RANS technique has failed, emphasis has changed to more intricate arrangements with flow characteristics. Particularly, corporate interest in applying LES to complicated engineering flows has been stimulated by decades of advancement in LES and the advent of cheap workstation clusters and massively parallel computers. Nevertheless, the RANS technique has not been substituted by LES, and it seems unlikely that it will be for some time due to two key factors, computational analytical tool for real-world engineering challenges firstly, notwithstanding the present computing ability for practical purposes, performing LES routinely still costs far too much in terms of computing; secondly, LES is not yet mature enough for users without sufficient results can be achieved with the level of solution accuracy that can be anticipated using experience and knowledge. Currently, LES has wide range of applications in turbulent flows, gas turbine, jet noise, aeroacoustics, and atmospheric science [1].

2.3 Future challenges

In the future, LES will be properly used for a wider series of flow problems and for more difficult problems with more multi-disciplinary uses. Nevertheless, numerous major issues/challenges related with LES and its applications such as wall layer modeling, SGS modeling, LES of turbulent combustion, generation methods for inflow boundary conditions, etc., are still existing. Before LES can become a trustworthy, strong engineering analysis tool that can be utilized as a substitute for RANS, there are still important problems that need to be overcome. It is extremely improbable that LES will entirely overtake RANS and become a design tool for the foreseeable future, without considerable years of LES experience [1].

Actually, the PBL is more complex as compared with LES for simulation of systems; however, this complexity is generated due to heterogeneous nature of the considering surface. For instance, the earth land surface is described by spatially varying elements, urban expansion, and undulating grounds, which can merge circulations and therefore alter the turbulence dynamics. These complex surfaces may significantly affect turbulence transport in many climate applications, for example, vegetation development, pollution, cloud formation, and storm formation. In summary, the LES is more employed than realistic PBL for multi-dimensional environmental flows.

Abbreviations

large eddy simulation	LES
Sub-grid scale	SGS
Reynolds-averaged Navier-Stokes	RANS
Direct numerical simulation	DNS
Turbulent kinetic energy	TKE
Planetary boundary layer	PBL
Digital filter generator	DFG
Periodic boundary condition	PBC

Author details

Aamir Shahzad*, Muhammad Kashif and Fazeelat Hanif
Modeling and Simulation Laboratory, Department of Physics, Government College University Faisalabad (GCUF), Faisalabad, Pakistan

*Address all correspondence to: aamirshahzad_8@hotmail.com

References

[1] Zhiyin Y. Large-eddy simulation: Past, present and the future. Chinese Journal of Aeronautics. 2015;**28**(1):11-24

[2] Menzies. Large eddy simulation applications in gas turbines. Philosophical Transactions of the Royal Society A. 2009;**367**:2827-2838

[3] Tucker P. The LES model's role in jet noise. Progress in Aerospace Sciences. 2008;**44**(6):427-436

[4] Meneveau C, Katz J. Scale-invariance and turbulence models for large-eddy simulation. Annual Review of Fluid Mechanics. 2000;**32**(1):1-32

[5] Domaradzki JA, Adams NA. Direct modelling of subgrid scales of turbulence in large eddy simulations. Journal of Turbulence. 2002;**3**(1):024

[6] Moeng CH, Sullivan PP. Large-eddy simulation. Encyclopedia of Atmospheric Sciences. 2015;**2**:232-240

[7] Veloudis I, Yang Z, McGuirk JJ, Page GJ, Spencer A. Novel implementation and assessment of a digital filter based approach for the generation of LES inlet conditions. Flow, Turbulence and Combustion. 2007;**79**(1):1-24

Chapter 2

An Effective $\mathcal{H}^2$-LU Preconditioner for Iterative Solution of MQS Integral-Based Formulation Problems

Salvatore Ventre, Bruno Carpentieri, Gaspare Giovinco, Antonello Tamburrino, Fabio Villone and Guglielmo Rubinacci

Abstract

We present iterative solution strategies for solving efficiently Magneto-Quasi-Static (MQS) problems expressed in terms of an integral formulation based on the electric vector potential. Integral formulations give rise to discrete models characterized by linear systems with dense coefficient matrices. Iterative Krylov subspace methods combined with fast compression techniques for the matrix-vector product operation are the only viable approach for treating large scale problems, such as those considered in this study. We propose a fully algebraic preconditioning technique built upon the theory of $\mathcal{H}^2$-matrix representation that can be applied to different integral operators and to changes in the geometry, only by tuning a few parameters. Numerical experiments show that the proposed methodology performs much better than the existing one in terms of ability to reduce the number of iterations of a Krylov subspace method, especially for fast transient analysis.

Keywords: $\mathcal{H}^2$-matrix, LU decomposition, MQS volume integral formulation, time domain, preconditioning, eddy currents, controlled thermonuclear fusion

1. Introduction

In this work, we consider the numerical solution of the Magneto-Quasi-Static (MQS) problem expressed in terms of an integral formulation based on the electric vector potential [1–3]. Integral formulations give rise to discrete models characterized by linear systems with dense coefficient matrices. Iterative Krylov subspace methods combined with fast compression techniques for the matrix-vector product operation are the only viable approach for treating large scale problems, such as those considered in this work. In this framework, a key role is played by the choice of the preconditioner, which has to be (i) robust enough to reduce significantly the number of iterations required to converge toward an accurate numerical solution, and (ii) efficient to reduce the overall computational and memory costs for the iterative solution. In this work, we focus on this critical computational aspect for the MQS problem. We propose a fully algebraic preconditioning technique built upon the theory of $\mathcal{H}^2$-matrix representation, which can be applied to different integral operators and to changes in the geometry only by tuning a few parameters.

Many algebraic preconditioning methods for solving dense linear systems arising from the discretization of integral operators are computed from a sparse approximation of the stiffness matrix, which retains the near-field mesh interactions giving rise to the most relevant contributions to the singular integrals of the model and is easy to be factorized or inverted [4–6]. However, our past experience in using such methods on our MQS problem, e.g., multilevel incomplete LU factorizations [7], sparse approximate inverses [8], and preconditioners computed from the sparse resistance matrix of the discrete system of equations (see Eq. 5 below) was not completely satisfactory. We found that the lack of global information due to the compact support of the sparse near-field matrix often results in slow convergence. For instance, in the CARIDDI code that implements the electric vector potential formulation of our problem [1, 9], a preconditioner based on the sparsity pattern arising from local interactions among neighboring basis functions is used to accelerate the iterative solution (see [10]). The aforementioned limitations are clearly shown in our numerical experiments, suggesting the need to develop global preconditioners that can incorporate contributions from the "far-field" interactions for this class of problems. In our model, far-field interactions can be included by taking into account the vector potential.

Prompted by these considerations, in this paper, we propose a class of preconditioners based on the so-called Hierarchical or $\mathcal{H}$-matrices [11], which provide a general mathematical framework for a highly compact and kernel independent accurate representation of integral equations with almost linear complexity [12]; see [13] for an example of application. $\mathcal{H}$ -matrices can be used to define global preconditioners built by means of the decomposition of small rank blocks at a moderate memory requirement. In previous works by other authors, highly efficient approximate LU-factorization and multilevel sparse approximate inverses preconditioners of boundary element matrices requiring an almost linear computational cost have been developed (see [14]) by means of the fast arithmetic of $\mathcal{H}$ -matrices. The novelty of this work is to use the latest development of $\mathcal{H}$ -matrices, that is, $\mathcal{H}^2$ -matrix theory [15, 16].

The paper is organized as follows. In Section 2, we present the MQS volume integral formulation used in our study and its numerical implementation. In Section 3, we establish the need for an effective preconditioner for our problem. In Section 4, we outline some fundamental properties of $\mathcal{H}^2$-matrix theory, and in Section 5, we describe the computational steps of a matrix factorization that has $\mathcal{O}(N \cdot \log(N))$ complexity, N being the number of DoFs (Degrees of Freedom) and is applied for the design of our preconditioner. We assess the effectiveness of the proposed method for eddy current computations in fusion devices simulations by numerical experiments in Section 6, also against other algebraic preconditioning strategies. Finally, in Section 7, we draw some concluding remarks arising from the study and some perspective for future work.

2. The volume integral numerical formulation

Here we recall the main assumptions at the basis of a Magneto-Quasi-Static volume integral formulation and of its numerical implementation. We refer to a conducting region V_C, without any topological constraint, characterized by the ohmic constitutive equation $\mathbf{J} = \sigma\mathbf{E}$, where $\mathbf{J}$ is the current density, σ is the material conductivity, and $\mathbf{E}$ is the electric field induced in the conductor by a set of time-varying external sources distributed with assigned current density $\mathbf{J}_S$.

The relevant equations in the magneto-quasi-stationary limit are

$$\mathbf{E} = -\frac{\partial \mathbf{A}}{\partial t} - \nabla\varphi \tag{1}$$

$$\nabla \times \mathbf{A} = \mathbf{B} \tag{2}$$

where

$$\mathbf{A}(\mathbf{r}, t) = \frac{\mu_0}{4\pi}\int_{V_c} \frac{\mathbf{J}(\mathbf{r}', t)}{|\mathbf{r} - \mathbf{r}'|} dV' + \mathbf{A}_s(\mathbf{r}, t) \tag{3}$$

and

$$\mathbf{A}_s(\mathbf{r}, t) = \frac{\mu_0}{4\pi}\int_{\mathbb{R}^3 - V_c} \frac{\mathbf{J}_s(\mathbf{r}', t)}{|\mathbf{r} - \mathbf{r}'|} dV'. \tag{4}$$

By imposing the Ohm law in weak form, we have

$$\int_{V_c} \mathbf{W} \cdot \left\{ \sigma^{-1}\mathbf{J} + \frac{\partial}{\partial t}\left[\frac{\mu_0}{4\pi}\int_{V_c} \frac{\mathbf{J}(\mathbf{r}', t)}{|\mathbf{r} - \mathbf{r}'|} dV'\right] + \frac{\partial \mathbf{A}_s}{\partial t} \right\} dV = 0 \tag{5}$$

for any $\mathbf{W} \in S$ and with $\mathbf{J} \in S$, S being the set of solenoidal vector functions with continuous normal component across V_C and zero normal component on the boundary of V_C.

The numerical model implemented in the CARIDDI code [1, 9] is obtained by applying the Galerkin method to Eq. (5), having expanded the unknown currents $\mathbf{J}$ as $\mathbf{J}(\mathbf{r}, t) = \sum_k I_k(t)\nabla \times \mathbf{N}_k(\mathbf{r})$, where $\mathbf{N}_k$ is the set of edge element basis functions whose uniqueness is assured by the tree-cotree gauge [17]. Upon discretization, we obtain the following linear dynamical system:

$$\begin{cases} \mathbf{L}\frac{d\mathbf{I}}{dt} + \mathbf{R}\mathbf{I} = -\frac{d\mathbf{V}_0}{dt}, \quad t \geq 0 \\ \mathbf{I}(0) = \mathbf{i}_0 \end{cases}, \tag{6}$$

where $\mathbf{I}(t) = [I_i(t)]$, $\mathbf{R} = [R_{ij}]$, $\mathbf{L} = [L_{ij}]$, $\mathbf{V}_0(t) = [V_{0,i}(t)]$ and

$$R_{ij} = \int_{V_c} \nabla \times \mathbf{N}_i(\mathbf{r}) \cdot \sigma^{-1}\nabla \times \mathbf{N}_j(\mathbf{r}) dV, \tag{7}$$

$$L_{ij} = \frac{\mu_0}{4\pi}\int_{V_c}\int_{V_c} \frac{\nabla \times \mathbf{N}_i(\mathbf{r}) \cdot \nabla \times \mathbf{N}_j(\mathbf{r}')}{|\mathbf{r} - \mathbf{r}'|} dV dV', \tag{8}$$

$$V_{0,k}(t) = \int_{V_c} \nabla \times \mathbf{N}_k(\mathbf{r}) \cdot \mathbf{A}_s(\mathbf{r}, t) dV. \tag{9}$$

The quantity $\mathbf{i}_0$ in (6) denotes the prescribed initial condition. The vector potential $\mathbf{A}_s(\mathbf{r}, t)$ is produced by a known source current density $\mathbf{J}_s(\mathbf{r}, t)$, and it can be computed by means of the Biot-Savart law. For the sake of simplicity, we assume that the source current density $\mathbf{J}_s$ is due to the current $i_s(t)$ flowing in a given coil.

The ODE in (6) is integrated in time by the following scheme [18]:

$$\begin{cases} \mathbf{A}\mathbf{I}_n = \mathbf{L}\mathbf{I}_{n-1} + \mathbf{V}_{0,n} - \mathbf{V}_{0,n-1} \\ \mathbf{I}_0 = \mathbf{i}_0 \end{cases}, \tag{10}$$

where $\mathbf{I}_n = \mathbf{I}(n\Delta t)$ and $\mathbf{V}_n = \mathbf{V}(n\Delta t)$, and matrix $\mathbf{A}$ is expressed as

$$\mathbf{A} = \mathbf{R}\Delta \mathrm{t} + \mathbf{L}. \tag{11}$$

The quantity Δt is the time integration step. Matrix $\mathbf{R}$ is sparse and local, whereas $\mathbf{L}$ is fully populated because it arises from the discretization of an integral operator. Note that $\mathbf{R}$, $\mathbf{L}$, and $\mathbf{A}$ are positive definite. The solution of large linear systems with a dense coefficient matrix poses a relevant computational challenge as it requires fast matrix solvers with reduced algorithmic and memory complexity. In this work, we solve Eq. (10) iteratively, for each prescribed n. To speed up the iterative solution, we develop an accurate and computationally efficient preconditioner $\mathbf{P}$ built upon the $\mathcal{H}^2$-matrix representation of matrix $\mathbf{A}$.

3. The need of an effective preconditioner

As it is well known, the main operations that are performed at each step of an iterative solver are: (i) the computation of the matrix-by-vector product $\mathbf{Ax}$, and (ii) the application of the preconditioner. The overall computational cost of the iterative solution is then proportional to the number of arithmetic operations required to carry out the multiplication $\mathbf{Ax}$ and to the number of iterations that are necessary to converge to a user-defined accuracy. The role of the preconditioner is to reduce the number of iterations and, ultimately, the overall solution cost [19]. In this work, we apply the preconditioner from the left and replace the linear system in (10) with the new equivalent system

$$\mathbf{P}^{-1}\mathbf{A}\mathbf{I}_n = \mathbf{P}^{-1}(\mathbf{L}\mathbf{I}_{n-1} + \mathbf{V}_{0,n} - \mathbf{V}_{0,n-1}), \tag{12}$$

where the preconditioned stiffness matrix $\mathbf{P}^{-1}\mathbf{A}$ is assumed to be well conditioned and to have most of its eigenvalues clustered around point one of the spectrum. Therefore, the "ideal" preconditioner has to be a good approximation to $\mathbf{A}^{-1}$. A critical issue in this context is the dependence of matrix $\mathbf{A}$ from Δt. In fast transient analysis, Δt has to be small enough to model properly the underlying dynamics. On the other hand, in slow transients Δt has to be large enough to cover the full-time interval of interest at acceptable computational cost. The dependence of $\mathbf{A}$ on Δt yields the dependence of $\mathbf{P}$ on Δt. According to (8), for small Δt the preconditioner should be tailored mainly on $\mathbf{L}$, whereas for large Δt the preconditioner should be tailored mostly on $\mathbf{R}$. Here and in the following, "small" and "large" Δt should be intended as compared with the slowest eigenmodes of the dynamical matrix $\mathbf{L}^{-1}\mathbf{R}$. In intermediate cases (Δt neither too small nor too large), the preconditioner should depend on both $\mathbf{R}$

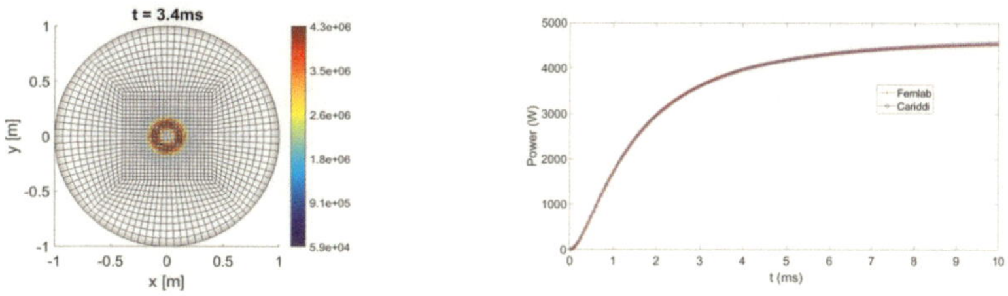

Figure 1.
Left: discretization of the plate used for assessing the performances of the preconditioner, together with the induced current distribution due to a current on a coaxial ring in a significant instant of the transient. Right: power losses on the plate as a function of time computed by CARIDDI (blue) and by Femlab© code (red), under the hypothesis of axisymmetry.

and **L**. Since the use of an implicit time stepping does not impose specific restrictions on Δt, the appropriate value of Δt depends on the underlying dynamics induced by the source, i.e., on the rate of change of $\mathbf{V}_0(t)$ defined in (9) and appearing in (6).

A canonical waveform for $i_s(t)$, which can be used as the elementary building block to approximate more complex waveforms[1], is given by a linear ramp $i_s(t) = \alpha t$, for $t \geq 0$, being α the rate of change for the electrical current. Consequently, $\mathbf{V}_0(t) = \mathbf{W}t$ for $t \geq 0$ is also linear. The source term $\mathbf{V}_{0,n} - \mathbf{V}_{0,n-1}$ is equal to $\mathbf{W}\Delta t$, which is constant for any n. Let us consider a transient starting from a vanishing initial condition $\mathbf{i}_0 = \mathbf{0}$ in (6). As long as the current at the previous step is negligible, Eq. (10) reduces to $\mathbf{A}\mathbf{I}_n = \mathbf{W}$. Therefore, the matrix to be inverted is **A**, and the preconditioner **P** has to be "tuned" on such matrix. On the other hand, when the stationary state is achieved, the solution $\mathbf{I}_n$ is constant, and Eq. (10) reduces to $(\Delta t\mathbf{R} + \mathbf{L})\mathbf{I}_\infty = \mathbf{L}\mathbf{I}_\infty + \mathbf{W}$, that is, $\Delta t\mathbf{R}\mathbf{I}_\infty = \mathbf{W}$. In this case, the matrix to be inverted is **R**, and the preconditioner needs to be "tuned" on **R**. This explains the reason why the preconditioner based on matrix **R** is not effective in dealing with fast dynamics. As an example, we measured its performances on a 3D transient eddy current problem where the driving current has a time-varying waveform given by $i_s(t) = \alpha t$, for $t \geq 0$, where $\alpha = 10^7 Hz$. This electrical current flows in a circular ring ($R = 10cm$), centered on the axis of a flat conducting disk (radius $R_0 = 1m$, thickness $3mm$). Validation is carried out against the Femlab© code [Version 3.51], as shown in **Figure 1** (right). **Figure 1**(left) shows the induced current distribution at $t = 3.4\ ms$ together with the finite element mesh used in the computation and the ohmic losses in the conductor. **Figure 2** shows the characteristics

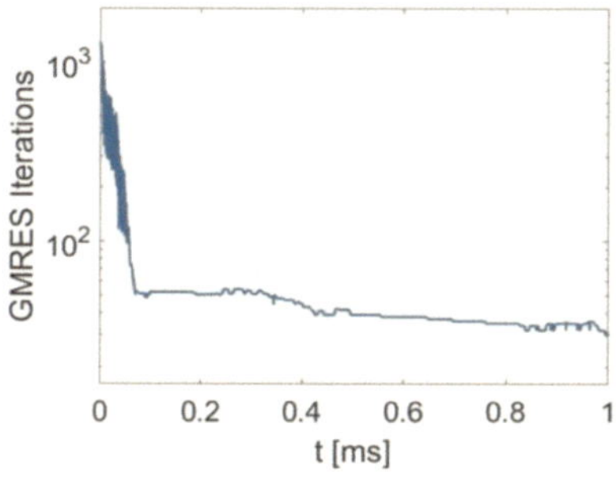

Figure 2.
Number of iterations of GMRES using R as a preconditioner during the transient.

[1] The linear ramp can be used to build a piecewise linear approximation of a continuous function.

of the iterative inversion using $\mathbf{P} = \mathbf{R}$ as a preconditioner and the Generalized Minimal Residual Method (GMRES) [20] as the iterative solver. Since the number of discrete unknowns for this test problem is 22,657, the results of **Figure 2** clearly show that that the resistive preconditioner ($\mathbf{P} = \mathbf{R}$) is not effective at the early stage of the transient phase, when the inductive effects accounted by matrix $\mathbf{L}$ are not negligible.

4. $\mathcal{H}^2$-matrix representation

The $\mathcal{H}^2$-matrix representation is widely used for storing efficiently a dense matrix arising from the discretization of a boundary integral equation [15]. In $\mathcal{H}^2$-matrix theory, the degrees of freedom (DoFs) of the underlying finite element mesh are partitioned into small subsets (or cluster) of nodes (they are edges in our problem). We illustrate the partitioning process with reference to the geometry shown in **Figure 3** that will be described in detail in the next section. At the first step, the set of DoFs is split in two clusters of nodes (see **Figure 3** (left)); then, each cluster is split recursively (see **Figure 3** (center) and (right)) until a minimum number of DoFs (called leaf size or LS) is reached by the recursive partitioning. The binary split is carried out using geometric information from the mesh. First, the coordinates of the barycenter of the DoFs are computed along the $x, y,$ and z axes. Then, the axis (either x or y or z) corresponding to the largest geometric coordinate in magnitude is selected, and the DoFs are split accordingly by introducing a plane orthogonal to this axis and passing through the median of the distribution of the coordinates of the DoFs along the same axis (see **Figure 4**). A key point of the partitioning is that cluster of nodes are built to contain DoFs that are geometrically neighbors in the finite element mesh. From a geometric viewpoint, all the DoFs belonging to a given cluster of edges are

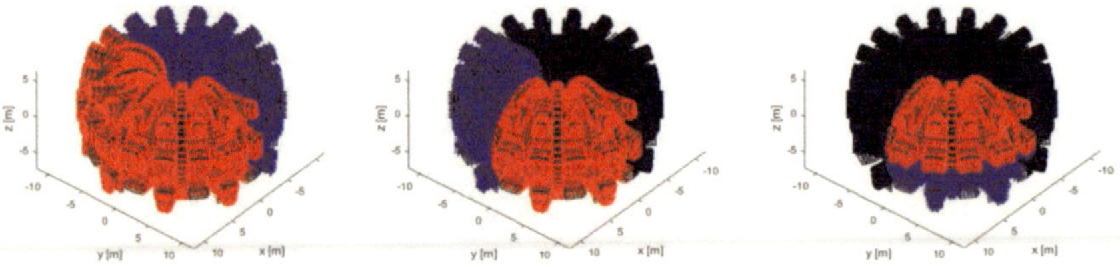

Figure 3.
The hierarchical partitioning of the DoFs. Left: The DoFs of the mesh are split into two sets (or clusters) of nodes. Center: The DoFs of the red set at the first step are partitioned in two subsets. Right: The DoFs of the red set at the second step are further partitioned into two subsets and so on.

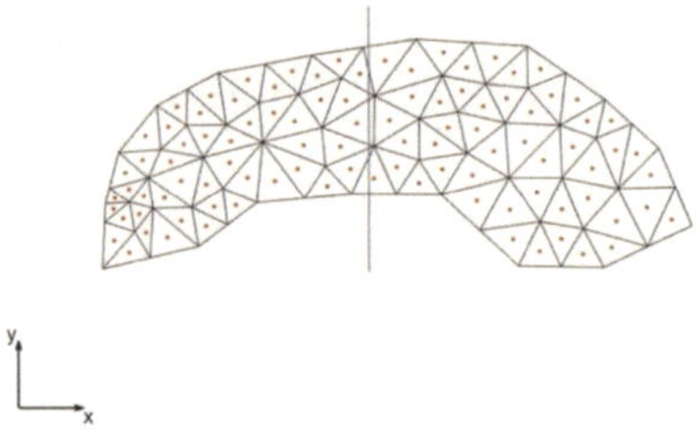

Figure 4.
Example of DOFs splitting for a two-dimensional (2D) unstructured mesh.

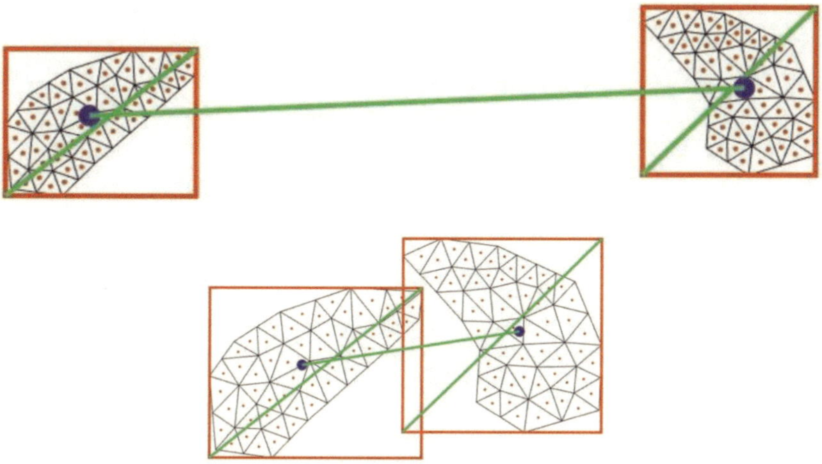

Figure 5.
Top: Two clusters that are classified as far. The size of the individual regions (given by the diagonal of the parallelepiped that bounds the region) occupied by the DoFs is larger than the distance between the barycenter of the clusters. Bottom: Two clusters that are classified as near. The size of the individual regions occupied by the DoFs is not large enough if compared with the distance between the barycenter of the clusters.

contained within a parallelepiped, which is parallel to the coordinate planes (see **Figure 5**). The key parameters related to a cluster N_k are: (i) the number of DoFs contained in the cluster, (ii) the parallelepiped P_k containing the DoFs of N_k, (iii) the diameter $d_k = |P_k|$ and the center c_k of P_k.

This partitioning process can be efficiently represented in a binary tree data structure. The binary tree is of paramount importance to classify interactions between clusters of DoFs in terms of near (or *inadmissible*) interactions and far (or *admissible*) interactions. An interaction is defined far if the two clusters of DoFs are geometrically well separated in the mesh, and near otherwise, as shown in **Figure 5**. We will use a simple criterion based on geometric distance for the classification. Two cluster of nodes N_k and N_j are named far if the ratio

$$\xi_{kj} \hat{=} \frac{|\mathbf{c}_k - \mathbf{c}_j|}{\max(d_k, d_j)}, \tag{13}$$

is greater than a suitable threshold $\xi_{THR} > 1$.

The mesh partitioning into clusters of nodes is represented by means of the so-called Block Cluster Tree (BCT) data structure (see **Figure 6** and [16] for details). Thanks to its binary structure, the BCT can be easily mapped into a corresponding partitioning of the discretization matrix $\mathbf{L}$ and, hence, of $\mathbf{A}$ into small blocks, yielding naturally a decomposition $\mathbf{A} = \mathbf{A}_{near} + \mathbf{A}_{far}$.

Interactions of near clusters lead to high or full-rank blocks (the near part $\mathbf{A}_{near}$ of $\mathbf{A}$) that need to be explicitely computed and stored, whereas far clusters interactions give rise to low-rank blocks (the far part $\mathbf{A}_{far}$ of $\mathbf{A}$). In the $\mathcal{H}^2$ -matrix representation, the far interactions between, say, the i-th and j-th clusters, are approximated by a low-rank matrix factorization of the form

$$\mathbf{A}_{ij} \approx \mathbf{V}_i \cdot \mathbf{S}_{ij} \cdot \mathbf{V}_j^{\mathrm{T}}. \tag{14}$$

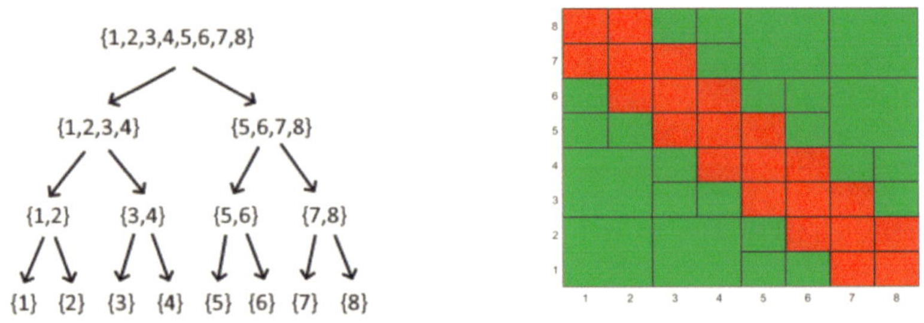

Figure 6.
Example of cluster tree partitioning ($n = 8$ clusters). Left: The block cluster tree. Right: Matrix representation of the block cluster tree after four levels of matrix block partitioning. Admissible blocks are green, and inadmissible blocks are red.

Such factorization can be obtained by using a degenerate Lagrange polynomial approximation of the kernel [3]. This is possible because the kernel defined in Eq. (8) is smooth when it is evaluated on the far field. The columns of matrix $\mathbf{V}_i$ represent the expansion basis for the DoFs associated to the i-th cluster, and $\mathbf{S}_{ij}$ is the coupling matrix of dimensions $k_i \times k_j$ where k_i and k_j are the column dimensions of $\mathbf{V}_i$ and $\mathbf{V}_j$, respectively. It is worth noting that for far interactions, integers k_i and k_j are expected to be much smaller than the number of rows of matrices $\mathbf{V}_i$ and $\mathbf{V}_j$. This representation of the far field is highly efficient in terms of both memory and computational costs, since it involves only small-rank matrices ($\mathbf{V}_i$ for any i, and $\mathbf{S}_{ij}$ for any i and j). Indeed, linear memory complexity can be achieved by means of the $\mathcal{H}^2$ -matrix representation [12].

Hereafter, the following notation will be used: we denote by N_p the number of Lagrange points per direction used for the degenerate kernel approximation, and by e_{ij} the relative error of the low-rank representation of a matrix block $\mathbf{A}_{ij}$ associated to the interactions of the i-th and j-th far clusters:

$$e_{ij} = \frac{\left\|\mathbf{A}_{ij} - \mathbf{V}_i \mathbf{S}_{ij} \mathbf{V}_j^{\mathrm{T}}\right\|}{\left\|\mathbf{A}_{ij}\right\|}, \tag{15}$$

where $\|\cdot\|$ is the conventional 2-norm for matrices.

5. The $\mathcal{H}^2$-matrix factorization

To develop an effective preconditioner $\mathbf{P}$ for solving Eq. (12), first we set $\mathbf{P}$ equal to $\mathbf{A}_{\mathcal{H}^2}$, the $\mathcal{H}^2$-matrix representation of $\mathbf{A}$. Then, by exploiting the hierarchical structure of $\mathbf{A}_{\mathcal{H}^2}$, we build an efficient LU factorization of $\mathbf{P}$. A core operation of the factorization of $\mathbf{P}$ is the computation of the Shur complement for the block matrix $\mathbf{A}_{\mathcal{H}^2}$. In general, this computation is very expensive. However, the $\mathcal{H}^2$-approximation introduces many off-diagonal blocks, which are vanishing upon a suitable orthonormal transformation of $\mathbf{A}_{\mathcal{H}^2}$. Hence, the overall factorization costs of $\mathbf{P}$ can be significantly reduced to $\mathcal{O}(N \cdot \log(N))$, N being the number of DoFs [16, 21].

The factorization algorithm proceeds level by level, starting from the last level of the BCT and computing a partial LU decomposition at each cluster of the tree. With reference to the leaf level, the last level in the BCT, the original stiffness matrix **P** is expressed as:

$$\mathbf{P} = \mathbf{L_o}\mathbf{C}_1\mathbf{U_p} \tag{16}$$

where $\mathbf{C}_1 = diag\left(\mathbf{I}_{d_1}, \mathbf{I}_{d_2}, \dots, \mathbf{I}_{d_{\#(leaf\ clusters)}}, \mathbf{C}\right)$ and $\mathbf{L_o}$ and $\mathbf{U_p}$ can be easily inverted because they are lower and upper triangular matrices, respectively. In (16), $\mathbf{I}_{d_i}$ is an identity matrix of proper size, and **C** is a full matrix of dimension much smaller than the original matrix **P**. After the factorization (16) has been computed at the current level, the algorithm continues at a higher level of the BCT by applying recursively the same process to matrix **C**.

Two key computational aspects have to be carefully considered and monitored during the factorization. The first one is the accuracy of the approximation $\mathbf{A}_{\mathcal{H}^2}$ to **A**, which is used to define the preconditioner **P**. Obviously, the closer $\mathbf{A}_{\mathcal{H}^2}$ to **A**, the more effective **P** is as a preconditioner for the iterative solver. The second aspect refers to the error introduced in the approximated factorization of **P**. In the rest of the section, we discuss in detail the factorization of **P**.

Specifically, matrix **P** is factorized in terms of a partial LU decomposition [16]. For doing this operation efficiently, using some properties of $\mathcal{H}^2$-matrix theory is of paramount importance to reduce the overall cost. The complete multilevel factorization algorithm is presented in Algorithm 1.

Algorithm 1: The iterative factorization algorithm.

---

1. Set $\mathbf{A}_{curr} = \mathbf{A}_{\mathcal{H}^2}$.
2. **for** k=last level to the root **do.**
3. **for** any cluster i at level k **do.**
4. Set $\mathbf{A}_{curr} = \mathbf{Q}_i^H\mathbf{A}_{curr}\mathbf{Q}_i$, where $\mathbf{Q}_i$ is defined in Eq. (14).
5. Compute the partial LU factorization of $\mathbf{A}_{curr}$ as.
   $\mathbf{A}_{curr} = \mathbf{L}_i\begin{bmatrix}\mathbf{I} & \mathbf{0}\\ \mathbf{0} & \tilde{\mathbf{A}}_{curr}\end{bmatrix}\mathbf{U}_i$, where $\mathbf{L}_i$ and $\mathbf{U}_i$ are lower and upper triangular matrices, respectively.
6. Set $\mathbf{A}_{curr} = \begin{bmatrix}\mathbf{I} & \mathbf{0}\\ \mathbf{0} & \tilde{\mathbf{A}}_{curr}\end{bmatrix}$
7. **end for.**
8. **end for.**

At step 5, matrix $\mathbf{Q}_i$ encompasses the basis related to the i-th cluster as follows:

$$\mathbf{Q}_i = diag\left\{\mathbf{I}_{d_1}, \mathbf{I}_{d_2}, \dots, \left[\mathbf{V}_i^{\perp}, \mathbf{V}_i\right], \dots, \mathbf{I}_{\#(leaf\ clusters)}\right\}. \tag{17}$$

The orthonormal transformation at step 5 is computationally cheap to carry out because of the simplified structure of matrix $\mathbf{Q}_i$, which involves only the i-th diagonal block of matrix $\mathbf{A}_{curr}$, of dimension $d_i \times d_i$. In the definition of $\mathbf{Q}_i$ in Eq. (17), $\mathbf{V}_i^{\perp}$ is the orthogonal complement of the cluster basis $\mathbf{V}_i$. This update of $\mathbf{A}_{curr}$, which essentially corresponds to a change of basis, is very effective on the far blocks to the i-th diagonal block of $\mathbf{A}_{curr}$. Thanks to (14), it is easy to verify that the first $d_i - k_\ell$ rows of these blocks are zeroed out, where k_ℓ is the rank used for the approximate factorization at the ℓ-th level of the BCT. Since most of the blocks in the original matrix $\mathbf{A}$ are far, they will be efficiently compressed by the procedure described above (see [16]). At step 10, a complete LU factorization of a dense matrix is required. However, this operation is rather cheap because matrix $\tilde{\mathbf{A}}_{curr}$, resulting from the previous steps, is small.

We highlight an important computational aspect of Algorithm 1. Upon processing the i-th cluster at step 5, the partial LU factorization of $\mathbf{A}_{curr}$ writes as

$$\mathbf{A}_{curr} = \begin{bmatrix} \mathbf{A}_{i'i'} & \mathbf{A}_{i'p} \\ \mathbf{A}_{pi'} & \mathbf{A}_{pp} \end{bmatrix} = \underbrace{\begin{bmatrix} \mathbf{L}_{i'i'} & 0 \\ \mathbf{A}_{pi'}\mathbf{U}_{i'i'}^{-1} & \mathbf{I} \end{bmatrix}}_{\mathbf{L}_i} \begin{bmatrix} \mathbf{I} & \mathbf{0} \\ \mathbf{0} & \tilde{\mathbf{A}}_{curr} \end{bmatrix} \underbrace{\begin{bmatrix} \mathbf{U}_{i'i'} & \mathbf{L}_{i'i'}^{-1}\mathbf{A}_{i'p} \\ \mathbf{A}_{pi'} & \mathbf{I} \end{bmatrix}}_{\mathbf{U}_i} \tag{18}$$

where $\mathbf{A}_{i'i'} = \mathbf{L}_{i'i'}\mathbf{U}_{i'i'}$, i' is equal to the number of nodes in the i-th clusters minus the approximation rank at the current level, and

$$\tilde{\mathbf{A}}_{curr} = \mathbf{A}_{pp} - \mathbf{A}_{pi'}\mathbf{U}_{i'i'}^{-1}\mathbf{L}_{i'i'}^{-1}\mathbf{A}_{i'p}. \tag{19}$$

Clearly, matrix $\tilde{\mathbf{A}}_{curr}$ still has a block structure, which is inherited from the cluster-to-cluster interactions. Contributions to the update term $\mathbf{A}_{pi'}\mathbf{U}_{i'i'}^{-1}\mathbf{L}_{i'i'}^{-1}\mathbf{A}_{i'p}$ due to near clusters mesh interactions are computed explicitely. On the other hand, contributions due to far clusters interactions are approximated using a low-rank factorization of the form given by Eq. (11), where cluster bases are simply updated by appending new columns pointing in the direction of the space orthogonal to the range of $\mathbf{V}_i$, thus avoiding to re-compute the whole basis from scratch. This approximate computation of the Schur complement is essential to achieve $\mathcal{O}(N \cdot \log(N))$ complexity. For further details, see [16].

Summing up, in this method two important sources of errors are introduced during the factorization:

- The far-field approximation. Rather than factorizing the original matrix $\mathbf{A}$, we factorize its $\mathcal{H}^2$-matrix representation $\mathbf{A}_{\mathcal{H}^2}$. The accuracy of this approximation depends on the number of Lagrange points (N_p) used to compute the $\mathcal{H}^2$-matrix approximation of $\mathbf{A}$.

- The error introduced in the Schur complement update (6) for each processed cluster. This error is controlled by a suitable threshold parameters and is the dominant one in the presented algorithm.

6. Numerical results and discussion

The numerical example described in this section refers to eddy current computations in fusion devices [22, 23]. The mesh of our case study is depicted in **Figure** 7.

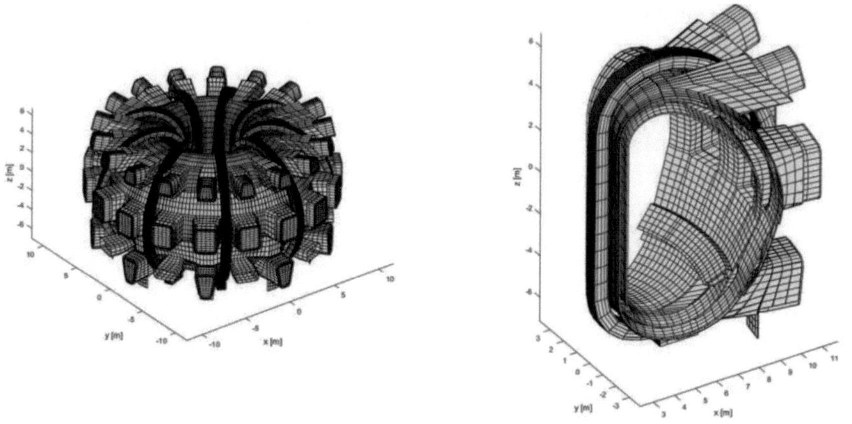

Figure 7.
The analyzed case study. Left: mesh of ITER magnet system. Right: A 45° sector.

The plasma made of hydrogen isotopes inside the vacuum chamber of the device can be broadly described as a deformable current-carrying conductor in the frame of the so-called Magneto-Hydro-Dynamics (MHD) equations. Consequently, the electromagnetic interaction of the plasma with surrounding conducting structures is fundamental in the description of the behavior of such devices. In nominal situations, when one wants to intentionally control plasma current, position, and shape [24], this interaction occurs on a time scale of the order of tens or hundreds of milliseconds in the largest existing devices. Conversely, also off-normal events can occur, such as the so-called disruptions [25]; in such cases, the plasma energy content is released on a much shorter time (milliseconds or submilliseconds time range).

It should be noted that the geometry of fusion devices, which is ideally axisymmetric, in practice can show significant three-dimensional features (ports, holes, cuts, slits, etc.), which must be taken into account when computing the magnetic fields, the current densities, and the related force densities, which are required to get a satisfactory electromagnetic modeling. Summing up, fusion devices are particularly challenging from the computational electromagnetic viewpoint: they are multiphysics devices, with complicated three-dimensional features giving rise to huge computational models, with different timescales of interests involved. The methods developed in the present paper are designed to enable us to tackle such intrinsic difficulties to some extent. The number of nodes and elements of the finite element mesh is 249849 and 132562, respectively, resulting in 145243 DoFs. The first three partitioning levels for such mesh are schematically depicted in **Figure 3**. In this study, the time step Δt is set equal to $1ms$ and $0.1ms$, to account for the different timescales mentioned above.

As aforementioned, the main operations required by an iterative solver are: (i) the computation of the matrix-by-vector product $\mathbf{Ax}$ and (ii) the application of the preconditioner at every iteration. We choose GMRES as the iterative solver. The preconditioner is the compressed approximate matrix $\mathbf{A}_{\mathcal{H}^2}$ presented in Section 5. Clearly, the use of a preconditioner introduces additional costs in terms of memory and computational cost, during the preprocessing and at each iteration. Therefore, it pays off only if there is a proper reduction of the number of required iterations. Here, to assess the performances of the preconditioner, we use matrix $\mathbf{A}$ without any approximation in computing the matrix-by-vector product during GMRES iterations. We stop the iterative process either when the initial preconditioned residual norm is reduced by five orders of magnitude or after 90 iterations. In our case, the computational time required to apply the preconditioner is proportional to its memory footprint. This is because of the almost linear complexity of the overall procedure (see [16] for details): both computational time and memory requirements increase linearly with the number of DoFs.

The performance of the proposed approach depends mainly on two key parameters: the number of Lagrange points N_p and the leaf size LS. We remind that LS determines the size of matrix $\mathbf{A}_{near}$ arising from near cluster interactions, whereas N_p affects the number of columns of $\mathbf{V}_i$, the size of $\mathbf{S}_{ij}$ and, consequently, the accuracy of $\mathbf{A}_{far}$. It is worth noting that $\mathbf{A}_{far}$ in the $\mathcal{H}^2$ -matrix representation does not contribute significantly to the overall required memory: in all numerical experiments of this work, the memory required by $\mathbf{A}_{far}$ is less than 10% of the storage required by $\mathbf{A}_{near}$.

From (11), we have that the preconditioner memory footprint depends only on the storage of matrices $\mathbf{L}_0$, $\mathbf{U}_p$ and the dense but smaller block $\mathbf{C}$ of matrix $\mathbf{C}_1$ appearing in (16). **Figure 8** shows the memory required by the preconditioner (M_{LU}) as a function of LS and N_p. The result is normalized with respect to the memory required to store the full matrix $\mathbf{A}$, which is about 169 *GB*.

The memory footprint required by matrix $\mathbf{C}_1$ decreases with LS. Indeed, an increase of LS enlarges the size of the block-to-block interactions, resulting in a smaller reduced matrix $\mathbf{C}_1$ because of a more effective compression of the block-to-block interactions. On the other hand, the sizes of $\mathbf{L}_0$ and $\mathbf{U}_p$ depend on: (i) the size of the near part, strictly depending on LS, and (ii) the ability of the $\mathcal{H}^2$ -matrix representation to compress the admissible block-to-block far-field interactions.

If we reduce LS, the memory required by the near interactions matrix $\mathbf{A}_{near}$ decreases, but the memory required by the far-field block-to-block interactions matrix increases. Indeed, we observe that at $LS = 500$ the required memory depends on N_p, whereas for $LS = 2000$ the required memory is almost independent on N_p. Eventually, we have an intermediate dependence for $LS = 1000$. The drawback related to the memory demands can be addressed in a parallel environment.

To assess the effectiveness of the proposed preconditioner, we make a comparison with the "reference" preconditioner based solely on matrix $\mathbf{R}$. This choice resulted to be much more effective than many other state-of-the-art algebraic preconditioners, such as robust multilevel, incomplete LU factorizations, and sparse approximate inverse methods to accelerate the GMRES convergence. Also, we evaluated a preconditioner based on the sum of matrix $\mathbf{R}$ plus a contribution from matrix $\mathbf{L}$ on the same pattern of $\mathbf{R}$. This preconditioner failed to converge at increasing mesh size.

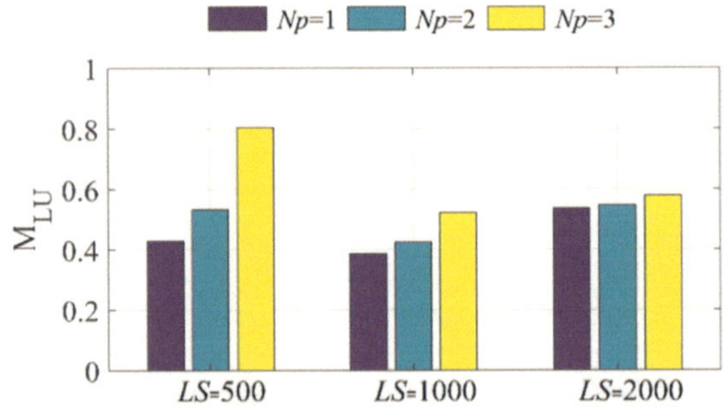

Figure 8.
Memory occupation as LS and N_p varies.

Summing up, finding a proper preconditioner for this class of large-scale problems is challenging.

Figure 9 (top $\Delta t = 1ms$) and (bottom $\Delta t = 0.1ms$) shows the preconditioned relative residual norm E_R of the GMRES solver versus the number of iterations for different LS and N_p.

The numerical results clearly show that:

- The $\mathcal{H}^2$ -LU generally performs much better than the preconditioner based on **R** preconditioner.

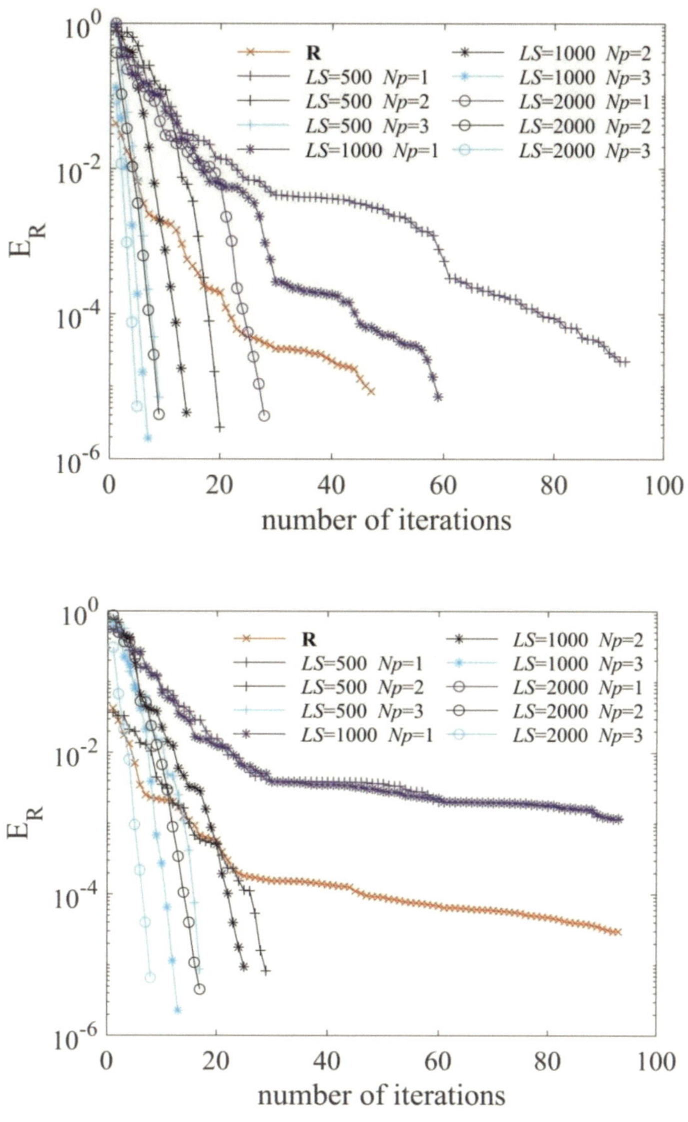

Figure 9.
Performance of $\mathcal{H}^2$ -LU preconditioner, for different values of the leaf size (LS) and number of Lagrange points (N_p). Top: $\Delta t = 1$ ms. bottom: $\Delta t = 0.1$ ms. performances from the reference preconditioner based on R are also shown.

- The efficiency of the preconditioner strongly depends on the choice of the parameters N_p and LS that control the accuracy of $\mathbf{A}_{far}$.

We highlight that the accuracy of $\mathbf{A}_{far}$ is rather low for $N_p = 1$ and that the preconditioner based on **R** performs better in this situation. In the other cases, the proposed $\mathcal{H}^2$ -LU preconditioner outperforms the reference one based on **R**, proving that it incorporates a good mechanism to balance contributions from matrices **L** and $\mathbf{R}\Delta t$ in **A**. This explains why at the smaller Δt, when **L** gives the dominant contribution to **A**, the number of iterations for the **R** based preconditioner increases. On the other hand, when using the $\mathcal{H}^2$ -LU preconditioner, which accounts for both $\mathbf{R}\Delta t$ and the dominant part of **L**, the number of iterations remains almost unchanged with Δt.

The accuracy of $\mathbf{A}_{far}$ is a critical point for the effectiveness of the proposed preconditioner. To asses this, we define two quantities, denoted as E_A and E_{SD}, that are the mean and the standard deviation for the relative error e_{ij} defined in (15) arising from all the admissible block-to-block pairs (i, j). As expected, **Figure 10** shows that the error decreases as N_p increases. Note that for $N_p = 1$, a relative error larger than 0.1 can be obtained. Eventually, the accuracy of the preconditioner is shown in **Figure 11** for different values of LS, N_p, and Δt. To quantify the error, we introduce the metric E_S defined as

$$E_S = \frac{\|I_{app} - I\|}{\|I\|}, \tag{20}$$

where I is computed using a direct method, whereas I_{app} is computed with the proposed preconditioned iterative solver. In (19), $\|\cdot\|$ is the 2-norm, and both I and I_{app} are evaluated at a prescribed time.

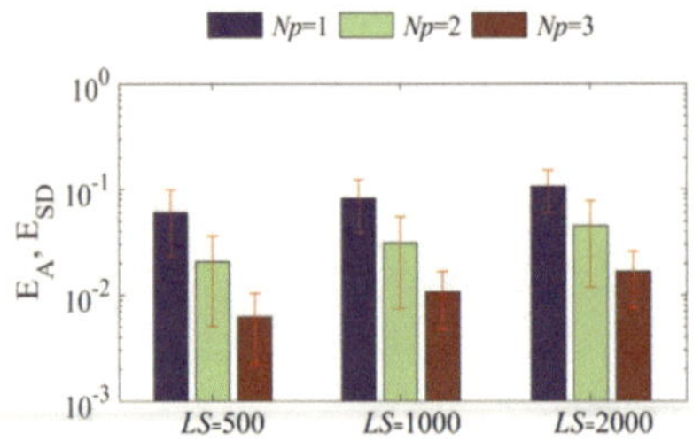

Figure 10.
Mean and standard deviation of the relative error in A_{far}.

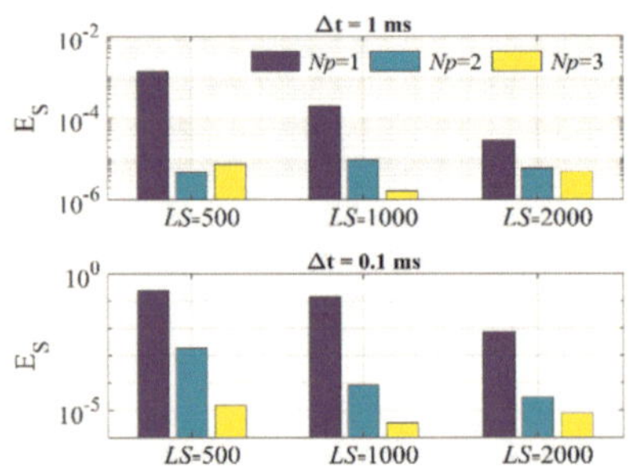

Figure 11.
Solution error E_s, as LS, N_p, and Δt vary.

Figure 11 confirms that the accuracy of the solution increases with N_p. Only for $LS = 500$ and $\Delta t = 1ms$ the solution with $N_p = 3$ is a little bit less accurate than the one with $N_p = 2$. It is worth noting that as the accuracy of $\mathbf{A}_{far}$ increases, the iterative scheme converges in very few iterations. Summing up, these numerical results confirm that an efficient iterative solver for the CARIDDI integral model of the eddy current problem needs a preconditioner incorporating some far-field approximations from the interaction matrix **L**.

7. Conclusions and future work

In this paper, a new preconditioner based on the $\mathcal{H}^2$ -LU decomposition has been introduced for solving integral equations models for eddy current problems. The application case refers to a large-scale model arising in nuclear fusion devices. The numerical results show that the preconditioner exhibits better performances compared with the purely resistive preconditioner. Specifically, the required number of iterations has been strongly reduced at different time steps. Moreover, the increase of costs in terms of both memory and CPU time during the preprocessing and at each iteration is largely paid off by the reduced number of iterations. In our view, two possible areas of future developments are: (i) deploying a parallel implementation of the $\mathcal{H}^2$ -LU decomposition in order to speed up the code execution and reduce the memory demands, and (ii) studying the performances of the method when it is used as a direct solver.

Acknowledgements

This work was partially supported by the program “Dipartimenti di Eccellenza 2018–2022,” MIUR, Italian Ministry of University and Research, by the project “CAR4PES,” 4th Marconi Fusion cycle, MUR PRIN Grant# 20177BZMAH and by the Research Südtirol/Alto Adige 2019 grant FH2ASTER, Provincia autonoma di Bolzano/Alto Adige – Ripartizione Innovazione, Ricerca, Università e Musei, contract nr. 19/34.

Author details

Salvatore Ventre[1], Bruno Carpentieri[2*], Gaspare Giovinco[3], Antonello Tamburrino[1,4], Fabio Villone[5] and Guglielmo Rubinacci[5]

1 Department of Electrical Engineering and Information, University of Cassino and Southern Lazio, Cassino, Italy

2 Faculty of Computer Science, Free University of Bozen-Bolzano, Bolzano, Italy

3 Department of Civil and Mechanical Engineering, University of Cassino and Southern Lazio, Cassino, Italy

4 Department of Electrical and Computer Engineering, Michigan State University, East Lansing, USA

5 Department of Electrical Engineering and Information Technology, University of Naples Federico II, Napoli, Italy

*Address all correspondence to: bruno.carpentieri@unibz.it

References

[1] Albanese R, Rubinacci G. Finite element methods for the solution of 3D eddy current problems. Advances in Imaging and Electron Physics. 1997;**102**: 1-86

[2] Rubinacci G, Tamburrino A, Villone F. Circuits/fields coupling and multiply connected domains in integral formulations. IEEE Transactions on Magnetics. 2002;**38**(2):581-584

[3] Rubinacci G, Tamburrino A. Automatic treatment of multiply connected regions in integral formulations. IEEE Transactions on Magnetics. 2010;**46**(8):2791-2794

[4] Carpentieri B. Algebraic preconditioners for the fast multipole method in electromagnetic scattering analysis from large structures: Trends and problems. Electronic Journal of Boundary Element. 2009;**7**(1):13-49

[5] Carpentieri B. Fast iterative solution methods in electromagnetic scattering. Progress in Electromagnetic Research. 2007;**79**:151-178

[6] Carpentieri B. A matrix-free two-grid preconditioner for boundary integral equations in electromagnetism. Computing. 2006;**77**(3):275-296

[7] Carpentieri B, Bollhöfer M. Symmetric inverse-based multilevel ILU preconditioning for solving dense complex non-hermitian systems in electromagnetics. Progress In Electromagnetics Research. 2012;**128**: 55-74

[8] Carpentieri B, Duff I, Giraud L, Sylvand G. Combining fast multipole techniques and an approximate inverse preconditioner for large electromagnetism calculations. SIAM Journal on Scientific Computing. 2005; **27**(3):774-792

[9] Albanese R, Rubinacci G. Solution of three dimensional eddy current problems by integral and differential methods. IEEE Transactions on Magnetics. 1988;**24**(1):98-101

[10] Rubinacci G, Ventre S, Villone F, Liu Y. A fast technique applied to the analysis of resistive wall modes with 3D conducting structures. Journal of Computational Physics. 2009;**228**(5): 1562-1572

[11] Hackbusch W. A sparse matrix arithmetic based on $\mathscr{H}$-matrices. Part i: Introduction to $\mathscr{H}$-matrices. Computing. 1999;**62**(2):89-108

[12] Börm S, Grasedyck L, Hackbusch W. Hierarchical Matrices. Leipzig: Max Planck Institute for Mathematics in the Sciences; 2003

[13] Voltolina D, Bettini P, Alotto P, Moro F, Torchio R. High-performance PEEC analysis of electromagnetic scatterers. IEEE Transactions on Magnetics. 2019;**55**(6):1-4

[14] Bebendorf M. Hierarchical LU decomposition-based preconditioners for BEM. Computing. 2005;**74**(3): 225-247

[15] Börm S. Efficient Numerical Methods for Non-Local Operators: $\mathscr{H}2$-Matrix Compression, Algorithms and Analysis. Vol. 14. Zürich, Switzerland: European Mathematical Society; 2010

[16] Ma M, Jiao D. Accuracy directly controlled fast direct solution of general $\mathscr{H}2$-matrices and its application to solving Electrodynamic volume integral equations. IEEE Transactions on

Microwave Theory and Techniques. 2017;**66**(1):35-48

[17] Albanese R, Rubinacci G. Integral formulation for 3D eddy-current computation using edge elements. IEE Proceedings A-Physical Science, Measurement and Instrumentation, Management and Education-Reviews. 1988;**135**(7):457-462

[18] Atkinson K, Han W, Stewart D. Numerical Solution of Ordinary Differential Equations. Vol. 108. Hoboken, New Jersey, U.S.: John Wiley & Sons; 2011

[19] Golub G, Loan CV. Matrix Computations: Johns Hopkins Studies in the Mathematical Sciences. third ed. Baltimore, MD, USA: The Johns Hopkins University Press; 1996

[20] Saad Y. Iterative Methods for Sparse Linear Systems. 2nd ed. Philadelphia: SIAM; 2003

[21] Ventre S, Carpentieri B, Giovinco G, Tamburrino A, Rubinacci G, Villone F. An $\mathcal{H}2$-LU preconditioner for MQS model based on integral formulation. In: Poster Presentation at the 22nd International Conference on the Computation of Electromagnetic Fields, COMPUMAG 2019, July 15–19. Paris, France; 2019

[22] Testoni P, Forte R, Portone A, Rubinacci G, Ventre S. Damping effect on the ITER vacuum vessel displacements during slow downward locked and rotating asymmetric vertical displacement events. Fusion Engineering and Design. 2018;**136**:265-269

[23] Ramogida G, Maddaluno G, Villone F, et al. First disruption studies and simulations in view of the development of the DEMO physics basis. Fusion Engineering and Design, Open Access. 2015;**96–97**:348-352

[24] Ariola M, Pironti A. Magnetic Control of Tokamak Plasmas. Vol. 187. Berlin/Heidelberg, Germany: Springer; 2008

[25] Hender T, Wesley J, Bialek J, et al. MHD stability, operational limits and disruptions. Nuclear Fusion. 2007;**47**(6): S128

Chapter 3

Large Scale Simulations for Dust Acoustic Waves in Weakly Coupled Dusty Plasmas

Aamir Shahzad, Zamar Ahmed, Muhammad Kashif, Amjad Sohail, Alina Manzoor, Fazeelat Hanif, Rabia Waris and Sirag Ahmed

Abstract

Dust acoustic wave of three-dimensional (3D) dusty plasmas (DPs) has been computed using equilibrium molecular dynamics (EMD) simulations for plasma parameters of Coulomb coupling strength (Γ) and Debye screening (κ). New simulations of wave properties such as longitudinal current correlation (LCC) $C_L(k, t)$ function have been investigated for 3D weakly DPs (WCDPs), for the first time. EMD results, C_L (k, t) have been simulated for four normalized wave numbers (k = 0, 1, 2, and 3). Our simulations illustrate that the frequency and amplitude of oscillation vary with increasing of Γ and κ. Moreover, present simulations of C_L (k, t) illustrate that the varying behavior has been observed for changing (Γ, κ) and system sizes (N). Current investigation illustrates that amplitude of wave oscillation increases with a decrease in Γ and N. However, there are slightly change in the value of C_L (k, t) and its fluctuation increases with an increasing k. The obtained outcomes have found to be more acceptable than those that of previous numerical, theoretical, and experimental data. EMD simulation has been performed with an increasing sequence for WCDPs and it serves to benchmark improved approach for future energy generation applications.

Keywords: plasma, complex (dusty) plasma, wave properties, weakly coupled (dusty) plasma, current correlation function, equilibrium molecular dynamics simulation

1. Introduction

Nowadays, it has been seen complex liquids have attracted considerable interest for many researchers. Complex liquids have introduced themselves by emerging the technology of all processes. Different techniques such as experimental, theoretical, and simulation are used to study the behavior of the complex liquid. Unambiguous models have been used to describe physical properties for specific ranges of temperature and pressure. Explicit equations can also be used to calculate the thermophysical

IntechOpen

properties of fluids when there is not any literature on complex fluids present. To study the behavior of complex fluids, information on thermophysical properties should be required. Transport properties of complex fluids play an essential role in laboratory and industrial applications. Moreover, thermophysical properties have played a very dominant role in optimization and system design. For the past few years, research on transport properties has occurred through several techniques such as experimental, theoretical, and simulation. Experimental research on transport properties has resulted in non-absolutely convergent expression because there is no accurate and precise experimental calculation of transport properties [1].

1.1 Plasma

Plasma makes up 99% of the matter in space and is referred to as the fourth state of matter. In essence, plasma is a sort of electrified gas in which atoms have split into positive ions and electrons. It is a type of matter in a variety of branches of physics, including astrophysics, technical plasma, and terrestrial plasma. Plasma is created artificially in a lab and used for a variety of technical applications, including fusion energy research, display, fluorescent lighting, and more. The term "plasma" was initially used by American scientist Irving Langmuir, who defined it as "a quasi-neutral gas of charged particles that exhibits collective behavior." When the number of ions equals the number of electrons ($n_i \approx n_e \approx n$), a gas is said to be "quasi-neutral," meaning it becomes electrically neutral. Where n is the number density, n_i is the ion density, and n_e is the electron density. The electric field and coulomb potential cause charged particles to collide when they exhibit collective behavior. The usage of plasma in research and technology is widespread. It is extremely important in daily life. Plasma is employed in many aspects of daily life, including lasers, sterilizing medical equipment, lightning, high-intensity beams, purifying water, and many more [2].

1.2 History of plasma

Plasma was initially defined in 1922 by American scientist Irving Langmuir, who was the only one to do so. Various academics began studying plasma physics in 1930 after being motivated by some particle puzzles. Hanes Alfven developed hydromagnetic waves in 1940; these waves are known as Alfven waves. He also mentioned that astrophysical plasma would be studied using these waves. Beginning around the same time in the Soviet Union, Britain, and the United States in 1950, research on magnetic fusion energy was initiated. The study of magnetic fusion energy was regarded as a subfield of thermonuclear power in 1958. This study was initially conducted under strict confidentiality, but after it became clear that the military did not appreciate controlled fusion research, the above-mentioned three countries made the research publicly available. Because of this, additional nations may take part in plasma-based fusion research. By the end of 1960, a Russian Tokomak configuration produces plasma with various plasma characteristics. Numerous sophisticated tokamaks that were created between 1970 and 1980 validated the effectiveness of the tokamak. Additionally, the Tokamak experiment nearly succeeded in achieving fusion break, and in 1990, work on DP physics started. When charged particles are absorbed by plasma, they split into four different components: electrons, ions, neutral particles, and dust particles and dust particles change plasma properties which are called as "Dusty plasma" [3].

1.3 Dusty plasma

When charged particles are submerged in plasma, the plasma properties become more complicated. This causes the plasma to be referred to as dusty plasma, which is also known as complex plasma. DP typically contains an extra charged particle and is electron-ion plasma. DPs contain dust particles of sizes ranging from tens of nanometers to hundreds of microns. The size of the particle is 3e-8g. This charged component can also be thought of as a micron-sized dust particle. Ice or metallic particles could make up dust particles. Different sizes and types of dust particles can be found throughout the universe and in the atmosphere. Although it typically takes the form of a solid, it can also be liquid or gaseous. The movement of ions and electrons can charge up dust particles. The electric potential of charged dust particles, which ranges from 1 to 10 V, is altered by the electric and magnetic fields. In a lab, dust particles can be grown. The existence of dust particles in plasma has exposed advanced fields for researchers and scientists. The majority of plasma in the universe is DP. It occurs frequently in the atmospheres of stars, the sun, planetary rings, galaxies, cosmic radiation, and the earth's magnetosphere and ionosphere. Based on the ordering of many radii and characteristic lengths between particles interacting (rd, λD), plasma with dust particles can be called either "dusty plasma" or "dust in plasma". If the $\lambda D > rd$ then it is called "dusty plasma" and if $\lambda D < rd$ then it is called "dust in plasma". Here, λD is dust particles' Debye length and rd is the interparticle distance. These are the conditions for DP [3, 4].

1.3.1 History of dusty plasma

Plasma is a very advanced field of science and technology. Many researchers have been motivated by their achievements in the field of science, medical field, and technology. In 1924, the term "plasma" first time defined by Irving Langmuir. Contemporary research on DP also has motivated many researchers and has become an interesting field of science and technology. In 1980, very exciting incident happened in the field of DP for the Saturn ring. In 1997, Mendis discovered a bright comet by a distant ancestor. The "distant ancestor" is an extraordinary comic laboratory for the investigation and study of interactions between dust particles and their physical and dynamic behaviors. Further manifestations of DPs were noctilucent clouds, origin nebula (which can be seen from the naked eye), zodiac light, etc. Images that are taken from the Saturn ring have shown that dust particles rotating around the Saturn ring have shape of spokes. The research from the last analysis has shown that these dust particles are fine particles. In 2000–2017, the present condition of DP is stable and is playing a main dominant role in science, industries, technology, energy sectors, and medical stores [5].

1.3.2 Charge on dust particle

Charging of dust grains can be drained through different processes involving the background plasma (electron and ion) being bombarded on surface of dust particles, secondary electron production, ion sputtering, photoelectronic emission by UV radiations, etc. Mostly, the charge on dust particles is negative in low-temperature laboratory plasma. When an electron from the background plasma strikes the surface of dust particle and is attached and lost from the plasma. Because electrons are more moveable than ions, the surface dust grain collects electrons, attracts ions, and repels electrons until the state of immobility is attained. This charge is accountable for a long lifetime of particles and confinement in plasma [4]. Other collective phenomena and wave instabilities are created due to interactions between these particles.

1.4 Types of dusty plasma

Dusty plasma is classified by the Coulomb coupling parameter.

- Weakly coupled plasma
- Strongly coupled plasma

1.4.1 Weakly coupled plasma

Weakly coupled plasma (WCP) is described by Coulomb coupling parameter Γ. Plasma is referred to as WCP when the coupling parameter's value becomes negligible. Hot plasma is another name for WCP. Plasma is referred to be hot plasma or ideal plasma when the temperature of the electron reaches the same level as the temperature of the ion (Te ≅Ti). In a laboratory discharge tube with high gas pressure, hot plasma is produced. The flame, atmospheric arcs, and sparks are examples of hot plasma. WCDP lacks a defined shape due to the extremely weak interactions between interacting particles at high temperatures and low densities. The EMD simulation has been used to examine the occurrence of diffusion motion in WCDPs [6].

1.4.2 Strongly coupled plasma

Strongly coupled plasma (SCP) is defined by the Yukawa potential, also known as the screened coulomb coupling potential (Γ). Coulomb coupling potential (Γ) is the ratio of potential energy to kinetic energy. Whenever potential energy exceeds unity, then kinetic energy becomes as $\Gamma > 1$. Its mean SCP is also defined as cold plasma since potential energy increases from inter-particle kinetic energy and plasma particles change into crystalline form. Cold plasma is used for teeth sterilization and food processing. Charged particles in SCP are influenced by the electric field, but the influence of the magnetic field is ignored for this kind of cold plasma.

1.5 Formation of dusty plasma in the laboratory

Different methods have been expanded for the production of DP in laboratories. Modified Q machines, rf discharges, and dc discharges are different methods that have been used in the last few years. Modified Q machine is single-ended machine used for the production of DP which allows the dust grains to dispersion over a cylindrical plasma column portion. In dc neon glow discharge stratum, DP is produced by a micron-sized dust particle. The discharge is created with cold electrodes in cylindrical glass tube. The distance between electrodes is 40 cm. The neon pressure is varied from 0.2 to 1 Torr and the fluctuation of the discharge current is 0.4–2.5 mA. DP is restrained in a cylindrical symmetric rf plasma system. The system of rf discharges consists of grounded electrodes, hollow outer electrode capacitive coupled to a 14 MHz rf power amplifier, and glass window. Man-made DP or DP generated in laboratory is dust in fusion devices, rocket exhaust, dust in space stations, dust precipitators, and thermonuclear fireballs. Other applications of DP are microelectronics fabrications; plasma enhanced chemical vapor deposition (PECVD), flat panel displays, solar cells, semiconductor chips, dusty plasma devices (DPDs) are used to produce DP in the laboratory. Ordinary flames, flame of candlelight, strong passion, fire, and blaze produced by burning gas are basically weakly ionized plasma that contains dust particles. Thermionic electron emission of 10 nm dust particles increases the degree of ionization in ordinary hydrocarbon flames. DP also contains volcanos and charged snow [7].

1.6 Acoustic modes

Acoustic modes are categorized into two different types such as collision-less, uniform, and unmagnetized dusty plasma. These are dust acoustic (DA) and ion-acoustic (IA) waves.

1.6.1 Dust acoustic (DA) wave

Dust Acoustic Wave (DAW) is a low-frequency type of longitudinal wave. Basically, DAW is a sound wave when it travels through complex fluid which causes the oscillation between charged particles. The mass of dust particles is considered very important because it provides inertia which is used to sustain DAW, pressure of ions and electrons affords the restoring force. The dust acoustic wave is examined in DP and presented the dispersion relation using the Boltzmann distribution of ion and electron density [8].

1.6.2 Ion acoustic (IA) wave

The propagation of sound waves in ordinary gas is absurd due to the absence of collision between particles. Thus, in ionized gas, there is no possibility of the occurrence of sound waves due to the absence of collision. However, in the case of plasma, ions can transmit vibrations to each other due to their charges. IA wave is a constant velocity wave. The occurrence of IA wave depends upon the thermal motion of ions. It is a longitudinal wave in magnetized plasma. It demonstrates reflection, diffraction, and interference phenomena. The phase velocity of electron thermal speed is much greater than the IA wave. Particle displacement, particle velocity, sound intensity, and sound pressure are essential quantities of IA waves.

2. Models and numerical simulations

Interactions of the dust particle are studied by many mathematical models such as "in the center of potential well and repulsion potential, in rigid sphere". There are a number of interaction potentials for the calculations of complex systems for instance DP. However, in the best of accurate experimental and theoretical balance, the Yukawa potential is more favorable interaction potential for WCDPs and is also used in many other physical systems (for instance, medicine and biology systems, chemicals, astrophysics environmental and physics of polymers, etc.) [9]. For charged particles, Yukawa potential has the following form [5]:

$$\phi(|r|) = \frac{Q^2}{4\pi\varepsilon_o} \frac{e^{\frac{-|r|}{\lambda_D}}}{|r|} \tag{1}$$

In the above equation, r represents inter-particle distance, Debye screening length is represented by λ_D, the permittivity of free space is represented by ε_0, and charge on dust particle is represented by Q. System particle is expressed by plasma parameters. Coulomb coupling strength (Γ) is defined as the potential energy of interaction between particles and the average kinetic energy of the particles. The Γ defines the distribution of plasma. Γ is given in this form of:

$$\Gamma = \left(\frac{Q^2}{4\pi\varepsilon_o} \frac{1}{a_{WS} k_B T} \right) \tag{2}$$

In Eq. (2), T is absolute temperature, k_B is Boltzmann constant, and a_{ws} represents Wigner Seitz (WS) radius and its value is $(3/4\pi n)^{1/3}$ with n representing number density ($n = N/V$) of dust particle. There is another parameter, k equal to $2\pi/L$ which affects the system of WCDPs. Where L represents the computational box length and $k^* = 2a_{ws}\pi/L$ is the normalized value of k. The $F_i = \sum_j F_{ij}$ is the force experienced on all dust particles. The force on ith particle is exerted by other particles. Debye screening strength is the ratio of the interparticle distance a (Wigner Seitz radius) to the Debye length λ_D [3, 5, 7, 10]. The value of κ can be represented in this form:

$$\kappa = \frac{a}{\lambda_D} \tag{3}$$

In this part, the EMD simulation is reported of C_L (k, t) of WCDP for large domain of plasma parameter of $(0.9 \leq \Gamma \leq 0.1)$ and $(1 \leq \kappa \leq 3)$ along with the varying value of wave number (k = 0, 1, 2 and 3). The accuracy and correctness of our proposed model for DAWs C_L (k, t) is checked with different system sizes (N = 500, 2048, and 4000).

For the system of 3D Yukawa system, time scales in WCDP are categorized by the inverse of plasma frequency as $(\omega_p{}^{-1})$ [1].

$$\omega_p^{-1} = \left(\frac{n Q_d^2}{\varepsilon_o m} \right)^{\frac{-1}{2}} \tag{4}$$

In Eq. (4), dust particle mass and number density are represented by m and n. In EMD simulation, the number of particles is N=500 to 4000 (the system size will be increased on the validity of computational power) placed in computational box.

2.1 Current correlation function

Local density provides information about the distribution of atoms. It is possible to study the local variations in motion of atoms. The momentum and particle current for single atomic particles are calculated by molecular dynamics simulation [9].

$$\pi(r, t) = \sum_j v_j \delta(r - r_j(t)) \tag{5}$$

With Fourier transformation

$$\pi(r, t) = \sum_j v_j e^{-ik.r_j(t)} \tag{6}$$

Fourier transform equation is used to measure the LCC function.

$$C_{\alpha\beta}(k, t) = \frac{k^2}{N_m} \left(\pi_\alpha(k,\ t) \pi_\beta(-k, 0) \right) \tag{7}$$

In Eq. (7), k is the wave vector. The equation given above can also be stated in form of the longitudinal and transverse current correlation function.

$$C_{\alpha\beta}(k, t) = \frac{k_\alpha k_\beta}{k^2} C_L(k, t) + \left(\delta_{\alpha\beta} - \frac{k_\alpha k_\beta}{k^2} \right) C_L(k, t) \tag{8}$$

Longitudinal and transverse current functions are acquired in X, Y, and Z directions by locating k are equal to $k\hat{z}$.

2.2 Normalized longitudinal current correlation function C_L (k, t)

Fourier transformation is used to write Eq. (7) into LCC functions which are normalized

$$C_L(k, t) = \frac{k^2}{N^m} \langle \pi_Z(k,\ t) \pi_Z(-k,\ \ 0) \rangle \tag{9}$$

By putting π_z value in Eq. (9), we obtain

$$C_L(k, t) = \frac{k^2}{N^m} \left\langle \sum v_j e^{-ik.Z_j(t)}(k,\ t) \sum v_j e^{-ik.Z}(k, t) \right\rangle \tag{10}$$

Eq. (10) explains charge particle transports in longitudinal motion. Longitudinal current correlation function describes that wave is propagating in negative and positive directions of the Z-axis.

2.3 Parameters and simulation techniques

This segment shows that system contains N = 500–4000 particles. Yukawa potential causes the interaction between these numbers of particles which is shown in Eq. (1) and contain in the cubic region. Moreover, this segment illustrates an overview of results attained through EMD simulation for longitudinal current correlation function which is shown in Eq. (10). The simulation takes place in simulation box and the simulation box dimension is selected as L_X, L_Y, and L_Z. In our case, we have used the EMD simulation to calculate the longitudinal current correlation function for N = 500 number of particles at plasma coupling parameter Γ = (0.1–0.9) with κ = (1.4, 2.0 and 3.0) at four varying values of wave numbers k = (0, 1, 2 and 3). The EMD simulation has been used to investigate shear viscosity and dynamical structure $S(k,\omega)$ of SCDPs [5, 7, 10–12].

3. EMD simulation results

3.1 Time-dependent longitudinal current correlation function C_L (k, t), at κ = 1.4

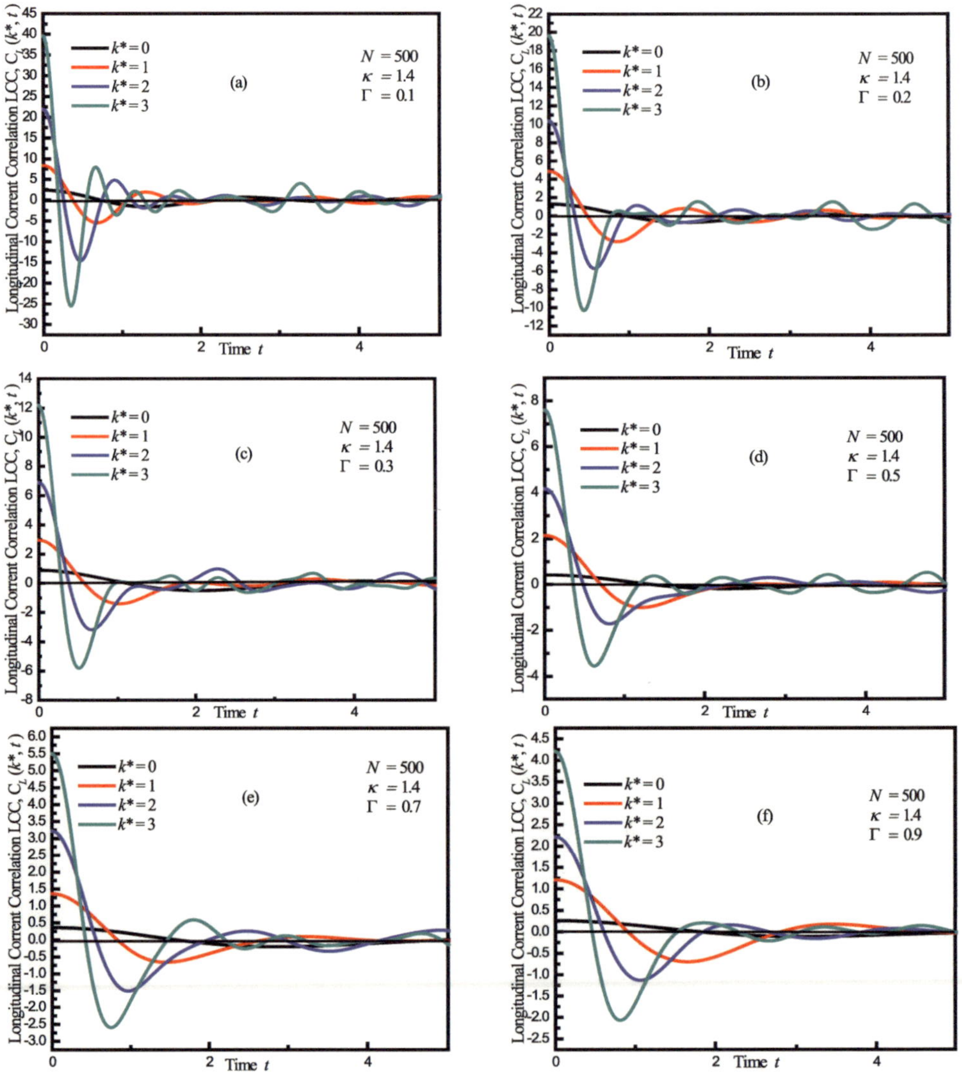

Figure 1.
The fluctuation of C_L (k, t) *as function of time of WCDP at system size* N = 500, κ = 1.4, *wave numbers* (k = 0, 1, 2 *and* 3) *for six different plasma parameters (a) Γ = 0.1, (b) Γ = 0.2, (c) Γ = 0.3, (d) Γ = 0.5, (e) Γ = 0.7, and (f) Γ = 0.9.*

3.2 Time-dependent longitudinal current correlation function C_L (k, t), at κ = 2.0

Computational results using EMD simulation have shown in **Figures 1** and **2** for weakly coupled Yukawa system LCC, *CL* (k, t) as a function of time at κ (= 1.4, 2.0 and 3.0) for six different values of Γ (= 0.1, 0.2, 0.3, 0.5, 0.7 and 0.9) and size of system is N=500 and four values of wavenumbers k (= 0, 1, 2 and 3) which are normalized. **Figures 1** and **2** explain main results for every three different values of

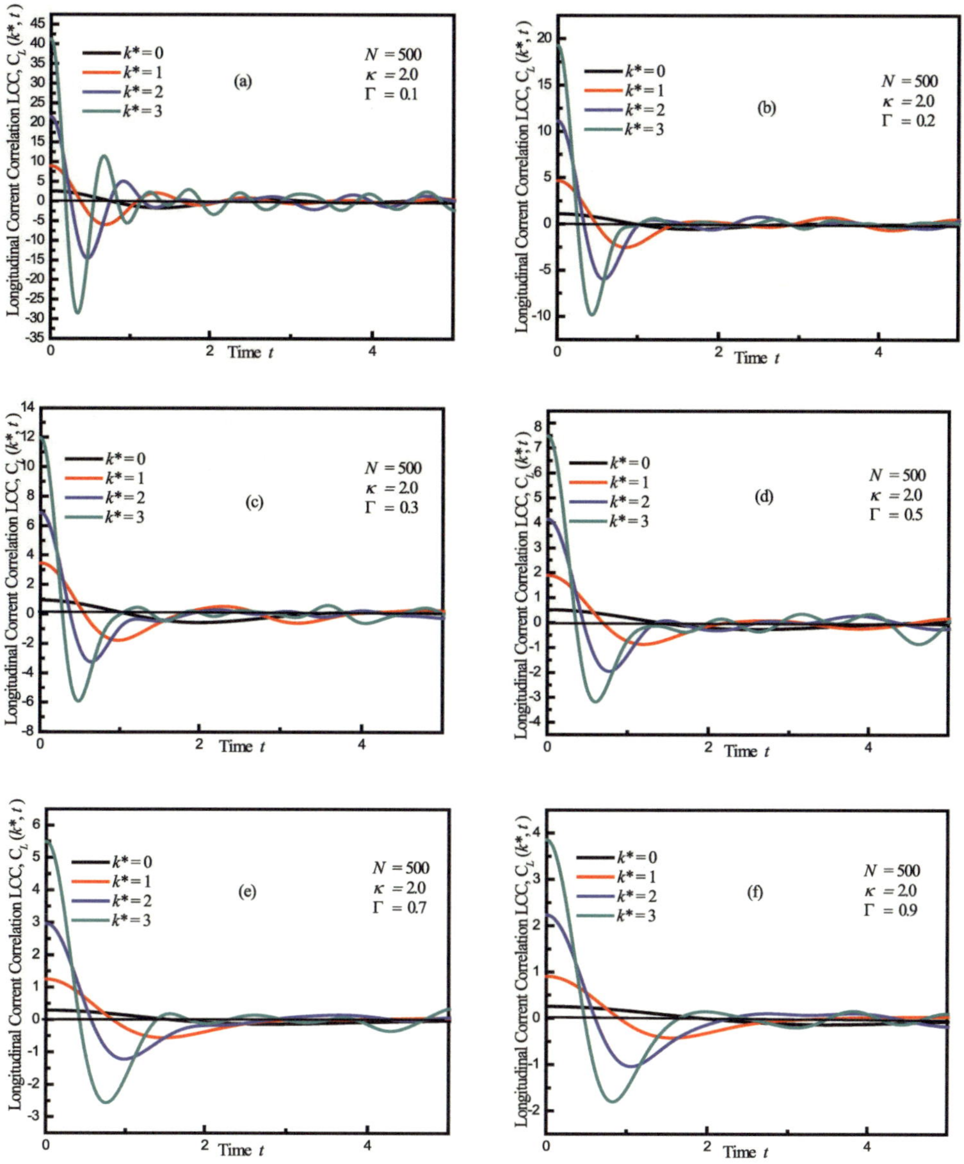

Figure 2.
The fluctuation of C_L (k, t) as function of time of WCDP at system size N = 500, κ = 2.0, wave numbers (k = 0, 1, 2 and 3) for six different plasma parameters (a) Γ = 0.1, (b) Γ = 0.2, (c) Γ = 0.3, (d) Γ = 0.5, (e) Γ = 0.7, and (f) Γ = 0.9.

the Debye screening parameter (κ = 1.4, 2.0, 3.0) of 3D Yukawa complex liquids. It is obtained from six panels of **Figure 1** that the amplitude and wavelength of LCC CL (k, t) increase with decrease of Γ (0.9–0.1). It is also obtained that by increasing wave numbers k (0, 1, 2, and 3), the amplitude of longitudinal current increases. It is noted that the CL (k, t) value for k = 2 and 3 shows maximum sinusoidal wave behavior at lower value of Γ from **Figures 1a–d** and **2a–c**. The sinusoidal wave behavior decreases with an increase of Γ and damping of the wave increases with an increase of time and Γ.

4. Summaries

Time-dependent LCC function, C_L (k, t) has been studied for weakly coupled Yukawa system through EMD simulation over a wider range of plasma parameters $(0.9 \leq \Gamma \leq 0.1)$, and $(1.4 \leq \kappa \leq 3.0)$ along with varying wave numbers values $(k = 0, 1, 2, 3)$. The behavior of the Weakly Coupled Yukawa system using EMD simulation has not been studied yet. It is observed that the amplitude and frequency C_L (k, t) of wave of oscillation increase when the value of Γ (0.9–0.1) decreases. C_L (k, t) value also increases when wave number k increases. The amplitude of C_L (k, t) steadily decreases with an increase in the number of particles N. It is implied that the presented simulation is based on Yukawa potential which has been employed to investigate the propagation of waves in WCDP. The presented simulation provides a great understanding of the propagation of the wave in WCDP. The employed simulation affords further respectable data for propagation in WCDP. It has been observed from these simulations that waves are repeatedly propagated at intermediate and higher values of Γ. The frequency and amplitude of waves are examined at different mode of waves.

Acknowledgements

This work was partially sponsored by the National Research Program for Universities (NRPU) with (No. 20-15251/NRPU/R&D/HEC/2021) for utilization of manpower resources. Moreover, this work also was partially supported by the industrial collaborative work between Government College University Faisalabad (GCUF) and Metal Industries Development Complex (MIDC) Sialkot, TEVTA, Govt. of Punjab, Pakistan, as a position of Consultant in a proposed project, for providing facilities to test and run our computer Experiment.

Abbreviations

EMD	equilibrium molecular dynamics
Γ	Coulomb coupling
κ	Debye screening strength
MD	molecular dynamics
PBCs	periodic boundary conditions
C_L (k, t)	longitudinal current correlation function
DP	dusty plasma
EMD	equilibrium MD
LCC	longitudinal current correlation
k	wave number
WCDPs	weakly coupled dusty plasmas
WCP	weakly coupled plasma
SCP	strongly coupled plasma
DA	dust acoustic
IA	ion acoustic
DAW	DUST ACOUSTIC WAVE

Author details

Aamir Shahzad*, Zamar Ahmed, Muhammad Kashif, Amjad Sohail, Alina Manzoor, Fazeelat Hanif, Rabia Waris and Sirag Ahmed
Modeling and Simulation Laboratory, Department of Physics, Government College University Faisalabad (GCUF), Faisalabad, Pakistan

*Address all correspondence to: aamirshahzad_8@hotmail.com; aamir.awan@gcuf.edu.pk

References

[1] Shahzad A, He MG. Thermal conductivity calculation of complex (dusty) plasmas. Physics of Plasmas. 2012;**19**(8):083707

[2] Chen FF. Introduction to Plasma Physics and Controlled Fusion. New York: Plenum Press; 1984. pp. 19-51

[3] Bellan PM. Fundamentals of Plasma Physics. New York, US: Cambridge University Press; 2008

[4] Rodriguez IJ. Some Assembly Required: Computational Simulations of Dusty Plasma. Portland, US: Portland State University; 2018. p. 622

[5] Shahzad A, Shakoori MA, Mao-Gang HE, Feng Y. Numerical Approach to Dynamical Structure Factor of Dusty Plasmas: Plasma Science and Technology-Basic Fundamentals and Modern Applications. Croatia/London, UK: INTECH Publisher; 2019

[6] Shahzad A, He M, Nikitin M. Thermal conductivity and non-newtonian behavior of complex plasma liquids. In: Thermoelectrics for Power Generation-A Look at Trends in the Technology. Rijeka: InTech; 2016

[7] Shukla PK, Mamun AA. Introduction to Dusty Plasma Physics. Florida, US: CRC Press; 2015

[8] Shukla PK. Twisted dust acoustic waves in dusty plasmas. Physics of Plasmas. 2012;**19**(8):083704

[9] Rapaport DC. The Art of Molecular Dynamics Simulation. New York, US: Cambridge University Press; 2004

[10] Shahzad A, He MG. Thermal conductivity of three-dimensional Yukawa liquids (Dusty Plasmas). Contributions to Plasma Physics. 2012; **52**(8):667-675

[11] Merlino R. Dusty plasma physics: Basic theory and experiments. Summer College on Plasma Physics. 2007;**30**: 1-130

[12] Xu W, Song B, Merlino RL, D'Angelo N. A dusty plasma device for producing extended, steady state, magnetized, dusty plasma columns. Review of Scientific Instruments. 1992; **63**(11):5266-5269

Chapter 4

Future Particle Accelerators

Javier Resta López

Abstract

Particle accelerators have enabled forefront research in high energy physics and other research areas for more than half a century. Accelerators have directly contributed to 26 Nobel Prizes in Physics since 1939 as well as another 20 Nobel Prizes in Chemistry, Medicine and Physics with X-rays. Although high energy physics has been the main driving force for the development of the particle accelerators, accelerator facilities have continually been expanding applications in many areas of research and technology. For instance, active areas of accelerator applications include radiotherapy to treat cancer, production of short-lived medical isotopes, synchrotron light sources, free-electron lasers, beam lithography for microcircuits, thin-film technology and radiation processing of food. Currently, the largest and most powerful accelerator is the Large Hadron Collider (LHC) at CERN, which accelerates protons to multi-TeV energies in a 27 km high-vacuum ring. To go beyond the maximum capabilities of the LHC, the next generation of circular and linear particle colliders under consideration, based on radiofrequency acceleration, will require multi-billion investment, kilometric infrastructure and massive power consumption. These factors pose serious challenges in an increasingly resource-limited world. Therefore, it is important to look for alternative and sustainable acceleration techniques. This chapter pays special attention to novel accelerator techniques to overcome present acceleration limitations towards more compact and cost-effective long-term future accelerators.

Keywords: high energy collider, Plasma wakefield, dielectric accelerator, solid-state plasma, plasmonic accelerator

1. Introduction

Currently, there are more than 30,000 particle accelerators in operation around the world, with a strong impact on science, industry and economy. Although, historically high energy physics (HEP) has been the main driving force for the accelerator R&D, only 1% of particle accelerators worldwide are used for research purposes, the rest are mostly used to support commercial, industrial and medical work. For instance, active areas of accelerator applications include radiotherapy to treat cancer, production of short-lived medical isotopes, synchrotron light sources, free-electron lasers, beam lithography for microcircuits, thin-film technology and radiation processing of food.

Since the pioneering work by Gustave Ising, Rolf Widerøe, Ernest Lawrence and many others in the 1920s and 1930s, the accelerator technology has progressed immensely [1]. Particularly, during the past 70 years, an impressive development of high energy colliders, with ever-increasing luminosities and centre-of-mass energies, has resulted in a number of fundamental discoveries in particle physics. In recent decades, experiments at hadron colliders, such as the Super Proton Synchrotron (SPS) and Tevatron, at e^+e^- colliders, such as the Large Electron-Positron Collider (LEP) and the Stanford Linear Collider (SLC), and at the electron-proton collider HERA (Hadron-Electron Ring Accelerator) have explored the energy range up to several 100 GeV and established beyond doubt the validity of the Standard Model in this range of energies [2]. In this tremendous progress, the development and optimisation of the radiofrequency (RF) technology has played a key role. Nowadays, most particle accelerators are driven by RF electromagnetic fields. Currently, the largest and most powerful RF-driven accelerator in operation is the Large Hadron Collider (LHC) at CERN [3], which accelerates protons to multi-TeV energies in a 27 km high-vacuum ring. Its kilometric size and complexity have earned it the name of the 'Cathedral' of the particle physics. With the discovery of the Higgs boson in 2012, the LHC has contributed to complete the Standard Model [4, 5].

To go beyond the maximum energy capabilities of the LHC, the next generation of RF-driven colliders under consideration would be among the largest and most complex facilities built on Earth, requiring intensive R&D and multi-billion Euro investments. For instance, the long-term goal of the Future Circular Collider (FCC) study hosted by CERN [6, 7] is the design and construction of a 100 TeV hadron collider in a 100 km long tunnel. Furthermore, future linear e^+e^- colliders based on RF technology, such as the International Linear Collider (ILC) [8] or the Compact Linear Collider (CLIC) [9], are designed to produce acceleration gradients of between 30 MV/m (ILC) and 150 MV/m (CLIC, 30 GHz frequency operation mode). These machines must, therefore, be tens of kilometres long to reach the desired beam energies, 125 GeV (ILC) and 1.5 TeV (CLIC). Another big project proposed by the HEP community is a high luminosity multi-TeV muon collider [10]. However, practically all the projects mentioned above, based on conventional RF acceleration technology, present serious challenges in terms of size and cost per GeV of beam energy. They need large-scale infrastructure (10–100 km scale) and, in most cases, a billion investment. To afford such a cost, it usually requires a consortium of several countries. Another key challenge is the sustainability in terms of the AC wall plug power consumption. High energy colliders are really 'hungry energy' machines that require hundreds of MW AC wall power to operate. To reduce such a power consumption in the latest years there have been a trend towards more 'green accelerators'. This concept includes the investigation of different ways to reduce energy consumption. For instance, through the improvement of the efficiency of existing technologies, e.g., increasing the efficiency of klystrons [11]. Also the design of Energy-Recovery Linacs (ERLs) can be considered as part of the R&D effort aimed at increasing energy efficiency [12].

To tackle the limitations of conventional accelerators, nowadays there is also an intensive R&D program on novel advanced concepts, some of which will be briefly described in the next sections. These alternative acceleration techniques might define the long-term future of particle accelerators, thus transforming the current paradigm in collider development towards more sustainable, compact and low-cost machines. For a more exhaustive review of past, present and future accelerators and colliders, see for example Refs. [2, 13, 14].

2. Plasma wakefield

Since the 1930s the RF acceleration technology has practically dictated the development of high energy colliders. Their power, cost and size have evolved with the

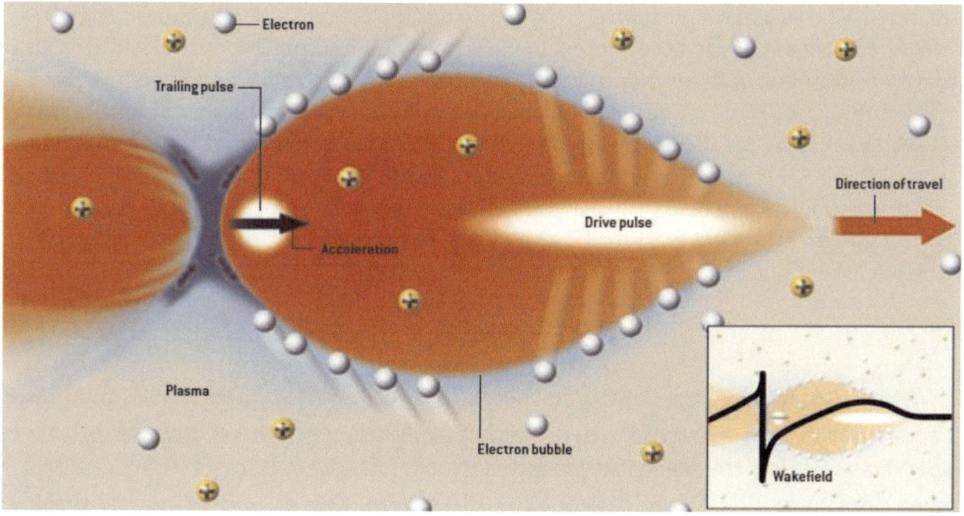

Figure 1.
Conceptual drawing of a plasma wakefield accelerator [17]. A drive pulse, either an intense laser (laser-driven) or a short electron bunch (beam driven) crossing a plasma, pushes away electrons. The much heavier ions are left behind. This charge separation makes a very large electric field. As the electrons rush back to their original position they overshoot forming a plasma wave, which induces a strong electric field behind the driver (wakefield). If a trailing charged particle pulse is positioned at the right spot seeing the right longitudinal wakefield phase behind the driver, it can be accelerated.

continuous improvements and optimisation experienced by RF cavities. However, over the recent decades, it has become apparent that RF technology is reaching its limits in terms of achievable accelerating gradients. Currently, the maximum electromagnetic fields that it can support are limited to approximately 100 MV/m due to surface breakdown [15]. To overcome this limit, Tajima and Dawson [16] proposed an alternative solution, a plasma accelerator based on laser-driven wakefields. Since plasma[1] is already a broken-down medium, there is no breakdown limit, compared to the conventional metallic RF cavities.

In general terms, the concept of Plasma Wakefield Acceleration (PWFA) is illustrated in **Figure 1**. A drive pulse enters the plasma and expels the electrons of the plasma outward. The plasma ions move a negligible amount due to their higher mass with respect to the electrons. In consequence, a positively charged ion channel is formed along the path of the drive pulse. The drive pulse can be either a short laser pulse (Laser Wakefield Acceleration [LWFA]) [16] or an electron bunch PWFA [18]. After the passage of the driver (laser or particle beam), the plasma electrons rush back in, attracted by the transverse restoring force of the ion channel. In this way a space charge-driven oscillation is excited, generating alternating regions of negative and positive charge, thus inducing a strong longitudinal electric field behind the driver, the so-called plasma wakefield. Therefore, if a witness charged particle bunch is injected behind the driver at the correct distance and phase, then it will be accelerated with high gradients, in some cases exceeding 100 GV/m.

In the last decades, acceleration of relativistic electrons in gaseous plasma has been successfully demonstrated in several experiments, see for example [19–21]. In the same way, PWFA of positrons has also been demonstrated [22–24]. Furthermore, the

---

[1] A plasma is defined as a fluid of positive and negative charges.

Advanced Wakefield Experiment (AWAKE) at CERN has recently shown that 10–20 MeV electrons can be accelerated to GeV energies in a plasma wakefield driven by a highly relativistic self-modulated proton bunch [25].

Previously, we have mentioned injection of a witness beam to be accelerated by the induced wakefield. In addition, internal injection of electrons from the plasma itself is also possible in the so-called 'bubble regime' [26–28].

The wakefields generated in plasma can be evaluated by analytical expressions [29]. The maximum accelerating field in a plasma accelerator can be estimated as follows:

$$E_z[\mathrm{V/m}] \approx 96\sqrt{n_e[\mathrm{cm}^{-3}]} \tag{1}$$

where n_e is the electron density in the plasma. This density determines the plasma frequency, $\omega_p = \sqrt{e^2 n_e / m_e \varepsilon_0}$, with m_e and e the electron rest mass and charge, respectively, and ε_0 the electrical permittivity of free space. For example, a typical plasma density $n_e = 10^{18}$ cm^{-3} gives $E_z \sim 100$ GV/m, that is, approximately three orders of magnitude higher than the maximum gradient obtained in RF structures.

Multi-TeV e^+e^- concepts based on staged plasma cells have been proposed [30–32], see **Figure 2**. However, one of the main drawbacks of plasma accelerators is the production of relatively poorer beam quality (higher energy spread and transverse size) with respect to conventional RF techniques. In order to tackle this problem and produce a reliable and competitive plasma-based accelerator technology, a big research consortium, EuPRAXIA, was recently established [33]. A comprehensive

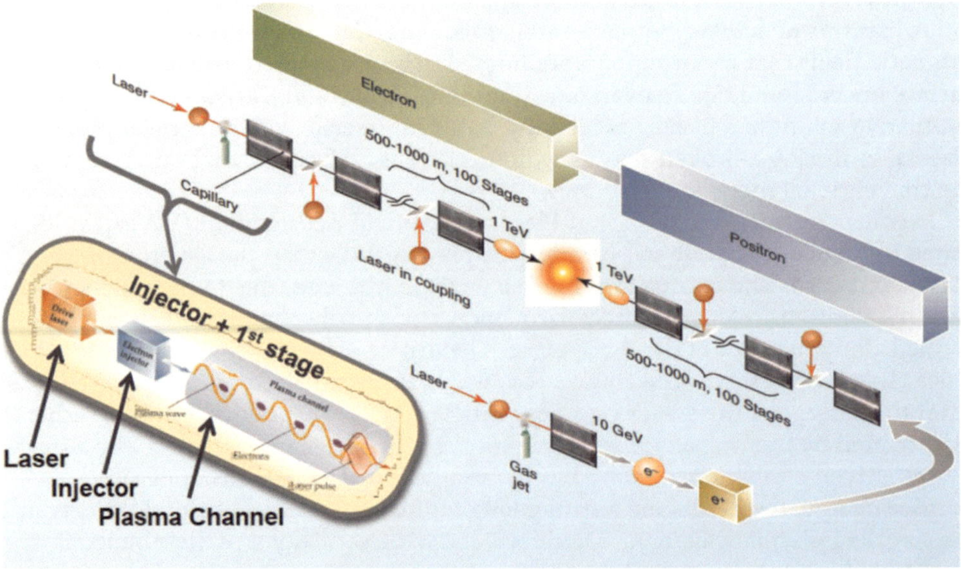

Figure 2.
Conceptual design of an LWFA-based linear collider. Both the electron and positron linacs consist of a sequence of plasma cells instead of RF cavities. The first stage consists of a plasma-based injection-acceleration module where controlled injection techniques are applied to produce a high-quality electron beam with approximately 10 GeV energy. Electrons are then accelerated to 1 TeV using 100 laser-plasma modules, each consisting of a 1 m long preformed plasma channel driven by a 30 J laser pulse giving a 10 GeV energy gain. A fresh laser pulse is injected into each module. Similarly, positrons are produced from a 10 GeV electron beam through pair creation and then trapped and accelerated in a laser-plasma acceleration module to approximately 10 GeV. Subsequent laser-plasma modules would accelerate positrons to 1 TeV [32].

review of near and long-term potential applications of plasma accelerators can be found in Ref. [34].

3. Solid-state-based acceleration

To go beyond the state-of-the-art, solid-state materials might offer new paths for beam manipulation and acceleration. For example, in the field of accelerator physics, the channelling properties of silicon crystals have successfully been used for collimation and extraction of relativistic proton beams [35]. Could also solids provide an alternative medium for acceleration? Depending on their particular atomic configuration and electrical conduction nature, some solid-state micro- and nano-sized structures offer interesting properties to enhance electric field components or induce strong wakefields that could be useful for acceleration, as well as transverse particle guiding and radiation emission. Next, we review some promising concepts that could revolutionise the future generation of accelerators and light sources.

3.1 Dielectric wakefield acceleration

Since the achievable peak field in the conventional RF metallic cavities is limited by surface breakdown, one obvious way to overcome this limit is through the use of materials with better breakdown properties, such as some dielectric materials (e.g., diamond, quartz, silica and ceramics). **Figure 3** illustrates the concept of beam-driven Dielectric Wakefield Acceleration (DWA) [36]. In this case, the dielectric accelerator consists of a hollow channel covered by a layer of a dielectric material and a metallic cladding. The electric field from the driver polarises the atoms of the dielectric medium, which coherently generates high frequency electromagnetic radiation.

Alternatively, lasers can be used as a driving source. Dielectric laser-driven acceleration (DLA) has been demonstrated for both non-relativistic [37] and relativistic electrons [38, 39] and has made enormous progress in recent years. DLA is the method behind the famous concept of ‘accelerator on a chip’ [40, 41]. DLA and DWA are limited to a maximum gradient of approximately 10 GV/m in the THz frequency range. **Figure 4** shows a particular dual-grating DLA configuration with sub-micron aperture.

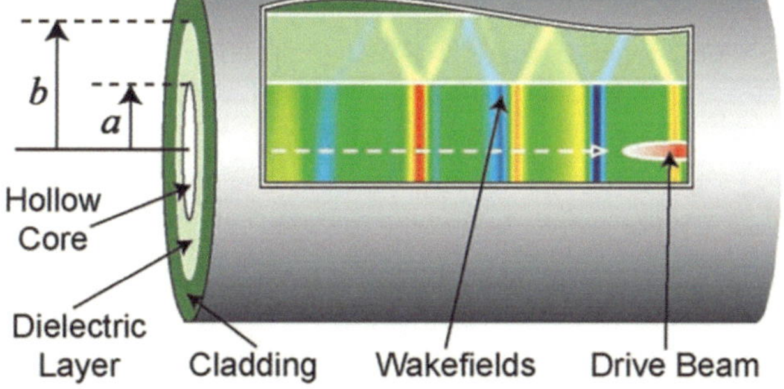

Figure 3.
Conceptual drawing of DWA [36]. A drive beam crosses a dielectric tube of radius “a“and excites wakefields. If a trailing beam is injected with the correct phase-space parameters, then it would be accelerated by the longitudinal component of the electric wakefield (bands of colour).

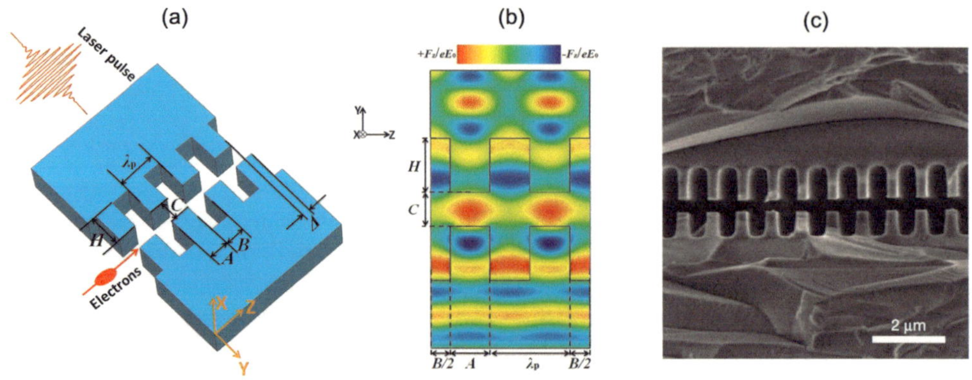

Figure 4.
(a) Conceptual scheme of a dual-grating scheme of DLA. (b) Simulation of the longitudinal accelerating force in a dual-grating structure illuminated by an input laser field along the y-axis [42]. (c) Scanning electron microscope image of the longitudinal cross-section of a DLA dual grating structure with 400 nm gap [38].

Usually, the dielectric structures have sub-mm dimensions, and there have been conceptual proposals to build a high-energy collider based on staged dielectric structures (**Figure 5**). This would allow 'table-top' high energy accelerators.

3.2 Crystal channelling acceleration

Semiconductor and metallic crystalline lattices have been proposed to generate a solid-state plasma medium to guide and accelerate positive particles, taking advantage of the channelling properties in crystals. High electron density in solids could be obtained from the conduction bands. Typical electron densities (n_e) in solid-state plasmas lie within the range of 10^{19} cm$^{-3} \leq n_e \leq 10^{24}$ cm^{-3} [43, 44], that is, between one and six orders of magnitude higher than the density in gaseous plasmas. Taking into account Eq. (1), solid-state-based plasmas might lead to accelerating gradients 0.1 TV/m$\lesssim E_z \lesssim$100 TV/m.

Solid-state wakefield acceleration using crystals was proposed in the 1980s and 1990s by T. Tajima and others [43, 45, 46] as a technique to sustain TV/m acceleration

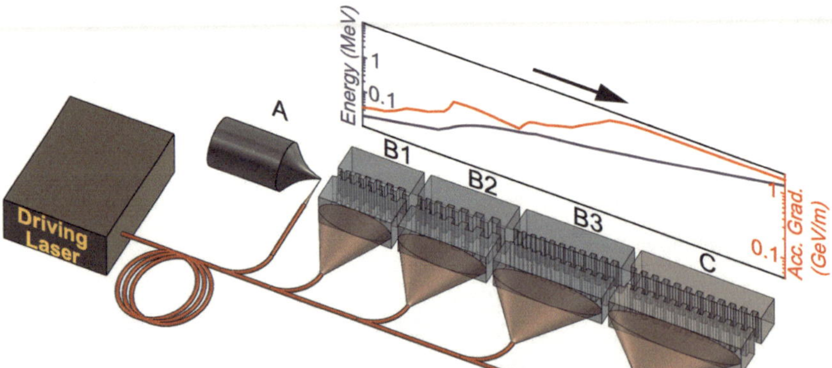

Figure 5.
Conceptual design of a staged DLA collider, consisting of dielectric structures with a tapered grating period to guarantee synchronicity with the accelerating electrons [37].

gradients. In the original Tajima's concept [45], high energy ($\simeq$40 keV) X-rays are injected into a crystalline lattice at the Bragg angle to cause the Borrmann-Campbell effect [47, 48], yielding slow-wave accelerating fields. Then a witness beam of charged particles, e.g., muons, which is injected into the crystal with an optimal injection angle for channelling, can experience acceleration along the crystal axis.

Wakefields in crystals can be induced by means of the excitation of high-frequency collective motion of conduction electrons through the crystalline lattice (see next section on plasmonic acceleration). To reach accelerating gradients on the order of TV/m, crystals must be excited by ultrashort X-ray laser pulses within a power range of TW– PW, which makes the practical realisation of the concept very challenging. It has only recently become a realistic possibility since the invention of the so-called single-cycled optical laser compression technique by G. Mourou et al. [49, 50].

If natural crystals (e.g., silicon) are used for solid-state wakefield acceleration, the beam intensity acceptance is significantly limited by the angstrom-size channels. In addition, such small size channels increase the dechannelling rate and make the channels physically vulnerable to high energy interactions, thus increasing the damage probability by high power beams.

Over the past decade, there have been great advances in nanofabrication techniques [51–53] that could offer an excellent way to overcome many of the limitations of natural crystals. Metallic nanostructures and metamaterials [54, 55] could lead to suitable ultra-dense plasma media for wakefield acceleration or charged particle beam manipulation, that is, channelling, bending, wiggling, etc. This also includes the possibility of investigating new paths towards ultra-compact X-ray sources [54]. The possibility to excite high acceleration gradients in multilayer graphene structures is also being investigated [56].

In the next section, we describe the plasmonic acceleration, which is a novel acceleration technique based on the excitation of suitable surface plasmonic modes in conducting materials. More concretely, we focus on plasmonic acceleration using nanostructures or metamaterials based on allotropes of carbon.

3.3 Plasmonic acceleration

Plasmonics can be defined as the study of the interaction between electromagnetic fields and the free electron Fermi gas in conducting solids. External electromagnetic fields can excite plasmons, that is, collective oscillations of conduction electrons in metals [57]. To some extent, this collective effect could be exploited to generate ultra-high acceleration gradients. **Figure 6** depicts a scheme of excited plasmons in metallic surfaces. The oscillation of induced longitudinal electric field reminds that in a sequence of a multi-cell RF cavity operating at π-mode.

The excitation of surface plasmonic modes [58, 59] can be driven either by charged particle beam [60] or by laser [61]. To be effective, the driver dimensions should match the spatial ($\sim$ nm) and time (sub-femtoseconds) scales of the excited plasmonic oscillations. Wakefield driving sources working on these scales are now experimentally realisable. For instance, attosecond X-ray lasers are possible thanks to the pulse compression technique [50]. In the case of beam-driven wakefields, the experimental facility FACET-II at SLAC [62] will allow the access to 'quasi-solid' electron bunches with densities up to 10^{24} cm^{-3} and submicron longitudinal size. A comparison of the range of spatial and time scales of the electric field oscillations and achievable acceleration gradient for standard and novel acceleration methods is shown

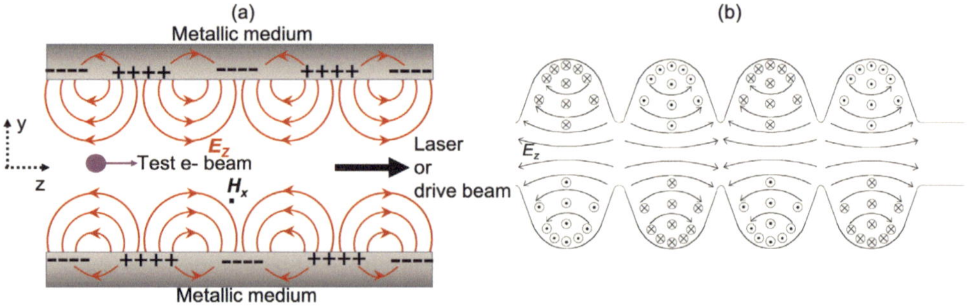

Figure 6.
(a) Plasmonic acceleration concept. (b) Comparison with an RF cavity operating in π-mode. The drawings are not to scale. While the plasmonic structure has micrometric or submicrometric apertures and lengths on the order of 1 mm, for instance, a 9-cell RF cavity has usually apertures of tens of mm and lengths on the order of m.

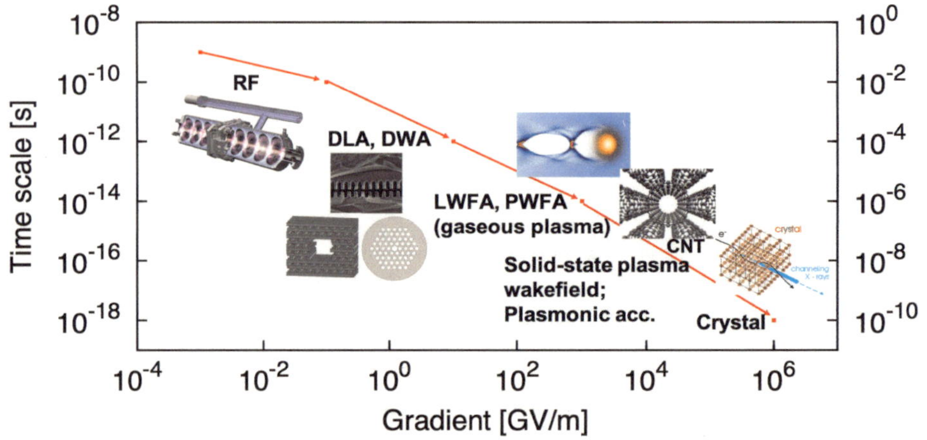

Figure 7.
Schematic comparing the space and time scales of the longitudinal electric field components generated by different techniques for charged particle acceleration and their corresponding amplitude (acceleration gradient). The cases of nanostructure wakefields, plasmonic and crystal acceleration are based on theoretical and numerical predictions.

in **Figure 7**. In principle, solid-state wakefields and plasmonics acceleration with nanostructures are predicted to have the potential to generate higher acceleration gradients than DWA and DLA, LWFA and PWFA with gaseous plasma. Plasmonic acceleration with nanostructures could be considered an intermediate step between PWFA and crystal channelling acceleration towards the PV/m gradient regime.

Due to their special thermo-mechanical and optoelectronic properties, materials based on carbon nanotubes arrays or graphene could offer an excellent medium to generate plasmonic wakefield acceleration. For instance, conduction electrons in CNTs could have densities $n_e \sim 10^{23}$ cm^{-3}. In principle, as proved in [63, 64], in the linear regime the plasmonic dynamics in CNT bundles can be described by classical plasma formulae. Therefore, from Eq. (1) we can estimate a maximum longitudinal wakefield $E_z \sim 10$ TV/m.

Particle-In-Cell (PIC) simulations of beam-driven wakefield acceleration in cylindrical metallic hollow structures (**Figure 8**) with micrometric or nanometric apertures have shown the feasibility of obtaining gradients $\gtrsim$100 GV/m, envisioning the possibility of an ultra-compact PeV linear collider [66–68]. Similar gradient values have

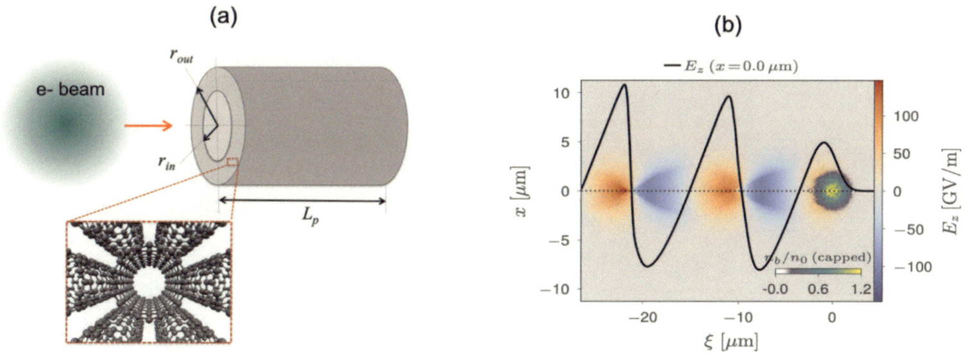

Figure 8.
(a) Schematic model for beam-driven wakefield simulation using a hollow cylinder of solid-state plasma confined in a wall of thickness $r_{out} - r_{in}$ and length L_p. In this model, the cylinder wall represents a solid made of CNT bundles. (b) an example of longitudinal wakefield as a function of the longitudinal coordinate $\xi = z - ct$, with c the speed of light. This particular case has been computed using the PIC code FBPIC [65].

also been predicted from PIC simulations of X-ray laser-driven wakefield acceleration in nanotubes [69].

One of the important properties of the plasmonic excitation is that the ionic lattice of the material remains strongly correlated. Unlike the plasmonic acceleration, other novel solid-state-based acceleration techniques require the complete ionisation of the ionic lattice or the ablation of solids, as described in the next section.

3.4 Target normal sheath acceleration

Solid-state plasmas can also be created by ablation of solids using high-intensity near-infrared femtosecond lasers. In this case, unlike in the plasmon wakefield excitation, the individual ions are uncorrelated. For example, the so-called target normal sheath acceleration (TNSA) technique is based on the ablation of a metallic thin foil [70]. The mechanism is schematically represented in **Figure 9**. A thin foil is irradiated by an intense laser pulse. The laser prepulse creates a pre-plasma on the target's front side. The main pulse interacts with the plasma and accelerates MeV-energy electrons mainly in the forward direction. The electrons propagate through the target, where collisions with the background material can increase the divergence of the electron current. Then the electrons leave the rear side, resulting in a dense sheath. An electric

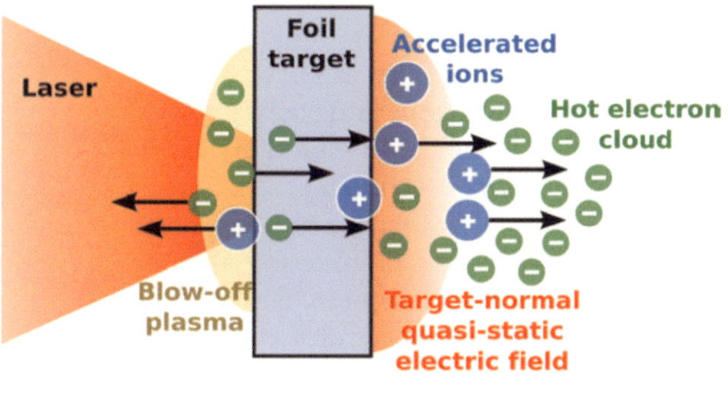

Figure 9.
Target normal sheath acceleration mechanism.

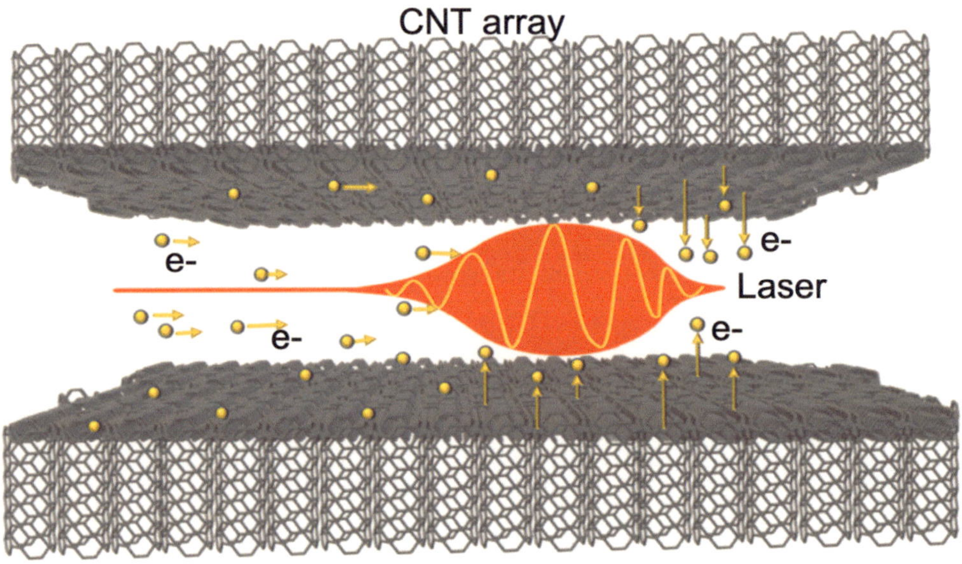

Figure 10.
Laser wakefield acceleration concept with aligned CNT arrays. In this particular configuration, CNTs are arranged perpendicularly to the laser propagation [71].

field due to charge separation is created. The field is of the order of the laser electric field (~TV/m), which ionises atoms at the surface. Eventually, the ions are accelerated in this sheath field, pointing in the target normal direction.

3.5 Solid-state plasma by electron field emission

Electron field emission from nanostructures could offer another interesting way to generate solid-state plasma. A possible configuration based on two layers of aligned CNT arrays and a gap in between them is shown in **Figure 10**. Taking into account the low work function of CNTs and their emission properties, a first intense laser pulse, travelling through the gap between the CNT sheets, could be used to induce electron field emission from CNT arrays. Then extracted electrons in the gap are accelerated by the ponderomotive force of a second laser pulse. For more details see Ref. [71].

4. Outlook and prospects

Currently, particle accelerators play an important role in many areas, including fundamental science, medicine, industry and society in general. During the last 100 years, the research in HEP has been the main driving force behind the development and progress of the accelerator technology. The LHC is currently the world's most advanced and powerful accelerator, and the only multi-TeV collider in operation. It consists of a 27-km ring of superconducting magnets with a number of accelerating structures based on RF electromagnetic fields to accelerate particles. The next generation of RF-driven high energy accelerators will pose serious challenges in terms of infrastructure size, technology and cost.

The current state-of-the-art of the RF techniques is limited to gradients on the order of 100 MV/m. Hence, larger and more expensive accelerator facilities are necessary in order to go beyond the LHC capabilities. Therefore, it is important to explore alternative advanced acceleration techniques to overcome the limitations of conventional technologies, thus progressing towards more compact, sustainable and economical particle accelerators. For many years, the development of inexpensive and compact 'table-top' high energy particle accelerators and light sources has been the ultimate dream of the accelerator physics community.

Some of the most promising novel acceleration techniques have briefly been reviewed in this chapter. For instance, PWFA and LWFA using gaseous plasma have experimentally demonstrated to be able to achieve $\sim$ 100 GV/m acceleration [19–24]. Another path towards TV/m acceleration is the use of special solid-state media, such as micrometric and millimetric grating dielectric structures, reaching $\sim$ 1–10 GV/m acceleration [38, 39].

Crystals and, more recently, nanostructures have also attracted attention to create a solid-state plasma medium with electron densities 1–6 orders of magnitude higher than those in gaseous plasmas. For example, recent theoretical and numerical studies have shown the feasibility of obtaining tens of TV/m fields in nanomaterials by excitation of non-linear plasmonic modes driven by a high-density sub-micron particle bunch [66, 67]. To obtain multi-TV/m fields other techniques rely on the ablation of solids [70] or the induced electron field emission from nanomaterials [71].

It is necessary to remark that the excitation of solid-state plasma wakefields requires the use of driving sources (lasers or charged particle bunches) at the same time and spatial scales as the wakefield oscillations in the plasma. In principle, wakefield driving sources working on solid-state plasma scales are now experimentally realisable. Attosecond X-ray lasers are possible thanks to the pulse compression [49, 50]. In the case of beam-driven wakefields, the experimental facility FACET-II at SLAC [62] could deliver 'quasi-solid' electron beams in the near future. Therefore, an experimental proof-of-concept of solid-state PWFA could possibly become a reality within the next 10 years.

It is also worth mentioning that nanostructures and metamaterials, due to their special and flexible optoelectronic properties, could also provide an excellent medium to create compact X-ray sources, e.g., X-ray light sources based on CNT arrays [72].

In conclusion, the different advanced acceleration techniques described in this chapter have the potential to transform the current paradigm in accelerator physics. They offer novel pathways to multi-GV/m and multi-TV/m field regimes. This will open new horizons to the physics of extreme fields, particularly in collider physics, light sources, and in many other areas of applied sciences, medicine and industry.

Acknowledgements

The author acknowledges support by the Generalitat Valenciana under grant agreement No. CIDEGENT/2019/058. This chapter is dedicated to the memory of Dr Esteban Fullana Torregrosa, a great physicist and best friend.

Abbreviations

AWAKE	Advance Wakefield Experiment
CERN	European Organisation for Nuclear Research
CLIC	Compact Linear Collider

CNT	Carbon Nanotube
DLA	Dielectric Laser-driven Wakefield Acceleration
DWA	Dielectric Wakefield Acceleration
ERL	Energy Recovery Linac
EuPRAXIA	European Plasma Research Accelerator with Excellence in Applications
FACET-II	Facility for Advanced Accelerator Experimental Test II
FCC	Future Circular Collider
HEP	High Energy Physics
HERA	Hadron-Electron Ring Accelerator
ILC	International Linear Collider
LEP	Electron-Positron Collider
LHC	Large Hadron Collider
LWFA	Laser Wakefield Acceleration
PIC	Particle-In-Cell
PWFA	Plasma Wakefield Acceleration
R&D	Research and Development
RF	Radiofrequency
SLAC	Stanford Linear Accelerator Center
SLC	Stanford Linear Collider
SPS	Super Proton Synchrotron
TNSA	Target Normal Sheath Acceleration

Author details

Javier Resta López
Institute of Materials Science (ICMUV), University of Valencia, Valencia, Spain

*Address all correspondence to: javier2.resta@uv.es

References

[1] Sessler A, Wilson E. Engines of Discovery. A Century of Particle Accelerators. Revised and Expanded Edition. Singapore: World Scientific; 2014. 280 p. DOI: 10.1142/8552

[2] Zimmermann F, Shiltsev VD. Modern and future colliders. Reviews of Modern Physics. 2021;**93**(1):015006. DOI: 10.1103/RevModPhys.93.015006

[3] Bruning O, Collier P. Building a behemoth. Nature. 2007;**448**:285-289. DOI: 10.1038/nature06077

[4] Collaboration ATLAS. Observation of a new particle in the search of the standard model Higgs boson with the ATLAS detector at the LHC. Physics Letters B. 2012;**716**:1-29. DOI: 10.1016/j.physletb.2012.08.020

[5] Collaboration CMS. Observation of a new boson at the mass of 125 GeV with the CMS experiment at the LHC. Physics Letters B. 2012;**716**:30-61. DOI: 10.1016/j.physletb.2012.08.021

[6] Benedikt M et al. FCC-ee: The Lepton collider. The European Physical Journal Special Topics. 2019;**228**:261-623. DOI: 10.1140/epjst/e2019-900045-4

[7] Benedikt M et al. FCC-hh: The hadron collider. The European Physical Journal Special Topics. 2019;**228**:755-1107. DOI: 10.1140/epjst/e2019-900087-0

[8] Michizono S. The international linear collider. Nature Reviews Physics. 2019;**1**:244-245. DOI: 10.1038/s42254-019-0044-4

[9] Stapnes S. The compact linear collider. Nature Reviews Physics. 2019;**1**:235-237. DOI: 10.1038/s42254-019-0051-5

[10] Long KR et al. Muon colliders to expand frontiers of particle physics. Nature Physics. 2021;**17**:289-292. DOI: 10.1038/s41567-020-01130-x

[11] Cai J, Syratchev I. KlyC: 1.5-D large-signal simulation code for klystrons. IEEE Transactions on Plasma Science. 2019;**47**(4):1734-1741. DOI: 10.1109/TPS.2019.2904125

[12] Litvinenko VN, Roser T, Chamizo-Llatas M. High-energy high-luminosity e^+e^- collider using energy-recovery linacs. Physics Letters B. 2020;**804**:135394. DOI: 10.1016/j.physletb.2020.135394

[13] Shiltsev VD. High energy particle colliders: Past 20 years, next 20 years and beyond. Physics-Uspekhi. 2012;**55**(10):965-976. DOI: 10.3367/UFNe.0182.201210d.1033

[14] Ferrario M. Present and future accelerator options beyond the LHC. Annalen der Physik. 2016;**528**(1–2):151-160. DOI: 10.1002/andp.201500240

[15] Abe T et al. Breakdown study based on direct in situ observation of inner surfaces of an rf accelerating cavity during a high-gradient test. Physical Review Accelerators and Beams. 2016;**19**:102001. DOI: 10.1103/PhysRevAccelBeams.19.102001

[16] Tajima T, Dawson JM. Laser Electron Accelerator. Physical Review Letters. 1979;**43**:267-270. DOI: 10.1103/PhysRevLett.43.267

[17] Joshi C. Plasma accelerators. Scientific American. 2006;**294**(2):40-47

[18] Chen P, Dawson JM, Huff RW, Katsouleas T. Acceleration of electrons by the interaction of a bunched Electron

beam with a plasma. Physical Review Letters. 1985;**54**(7):693-696. DOI: 10.1103/PhysRevLett.54.693

[19] Gordon D et al. Observation of Electron energies beyond the linear dephasing limit from a laser-excited relativistic plasma wave. Physical Review Letters. 1998;**80**:2133-2136. DOI: 10.1103/PhysRevLett.80.2133

[20] Leemans WP et al. Multi-GeV Electron beams from capillary-discharge-guided Subpetawatt laser pulses in the self-trapping regime. Physical Review Letters. 2014;**113**(24): 245002. DOI: 10.1103/PhysRevLett.113.245002

[21] Litos M et al. High-efficiency acceleration of an electron beam in a plasma wakefield accelerator. Nature. 2015;**515**:92-95. DOI: 10.1038/nature13882

[22] Blue BE et al. Plasma-Wakefield acceleration of an intense positron beam. Physical Review Letters. 2003;**90**: 214801. DOI: 10.1103/PhysRevLett.90.214801

[23] Corde S et al. Multi-gigaelectronvolt acceleration of positrons in a self-loaded plasma wakefield. Nature. 2015;**524**: 442-445. DOI: 10.1038/nature14890

[24] Doche A et al. Acceleration of a trailing positron bunch in a plasma wakefield accelerator. Scientific Reports. 2017;**7**:14180. DOI: 10.1038/s41598-017-14524-4

[25] Adli E, et al. (AWAKE Collaboration). Acceleration of electrons in the plasma wakefield of a proton bunch. Nature 2018;561:363–367. DOI: 10.1038/s41586-018-0485-4.

[26] Geddes CGR et al. High-quality Electron beams from a laser wakefield accelerator using Plasma-Channel guiding. Nature. 2004;**431**(7008): 538-541. DOI: 10.1038/nature02900

[27] Mangles SPD et al. Monoenergetic beams of relativistic electrons from intense-laser plasma interactions. Nature. 2004;**431**(7008):535-538. DOI: 10.1038/nature02939

[28] Faure J et al. A laser-plasma accelerator producing Monoenergetic Electron beams. Nature. 2004;**431**(7008): 541-544. DOI: 10.1038/nature02963

[29] Esarey E, Schroeder CB, Leemans WP. Physics of laser-driven plasma-based electron accelerators. Reviews of Modern Physics. 2009;**81**(3): 1229-1285. DOI: 10.1103/RevModPhys.81.1229

[30] Leemans WP, Esarey E. Laser-driven plasma-wave electron accelerators. Physics Today. 2009;**62**: 44-49. DOI: 10.1063/1.3099645

[31] Schroeder CB, Esarey E, Geddes C, Benedetti C, Leemans W. Physics considerations for laser-plasma linear colliders. Physical Review Special Topics-Accelerators and Beams. 2010;**13**:101301. DOI: 10.1103/PhysRevSTAB. 13.101301

[32] Leemans W. White paper of the ICFA-ICUIL, joint taks force: High power laser technology for accelerators. In: ICFA Beam Dyn. Newslett. Vol. 56. International Committee for Future Accelerators (ICFA). 2011. pp. 10-88. Available from: https://icfausa.jlab.org/archive/newsletter/icfa-bd-nl-56.pdf

[33] Walker PA et al. Facility considerations for a European plasma accelerator infrastructure (EuPRAXIA). In: Proceedings of the 10th International Particle Accelerator Conference (IPAC'19); 19-24 May 2019; Melbourne, Australia: Paper THPGW025.

[34] Joshi C, Corde S, Mori WB. Perspectives on the generation of electron beams from plasma-based accelerators and their near and long term applications. Physics of Plasmas. 2020; **27**:070602. DOI: 10.1063/5.0004039

[35] Scandale W, Taratin AM. Channeling and volume reflection of high-energy charged particles in short bent crystals. Crystal assisted collimation of the accelerator beam halo. Physics Reports. 2019;**815**:1-107. DOI: 10.1016/j.physrep.2019.04.003

[36] Thompson MC. Breakdown limits on Gigavolt-per-meter Electron-beam-driven Wakefields in dielectric structures. Physical Review Letters. 2008;**100**:214801. DOI: 10.1103/PhysRevLett.100.214801

[37] Breuer J, Hommelhoff P. Laser-based acceleration of nonrelativistic electrons at a dielectric structure. Physical Review Letters. 2013;**111**:134803. DOI: 10.1103/PhysRevLett.111.134803

[38] Peralta EA et al. Demonstration of electron acceleration in a laser driven dielectric microstructure. Nature. 2013; **503**:91-94. DOI: 10.1038/nature12664

[39] O'Shea BD et al. Observation of acceleration and deceleration in gigaelectron-volt-per-metre gradient dielectric wakefield accelerators. Nature Communications. 2016;7:12763. DOI: 10.1038/ncomms12763

[40] Blau SK. Particle acceleration on a chip. Physics Today. 2013;**66**(12):19. DOI: 10.1063/PT.3.2205

[41] England RJ, Hommelhoff P, Byer RL. Microchip accelerators. Physics Today. 2021;**74**(8):42-49. DOI: 10.1063/PT.3.4815

[42] Wei Y et al. Beam quality study for a grating-based dielectric laser-driven accelerator. Physics of Plasmas. 2017;**24**: 023102. DOI: 10.1063/1.4975080

[43] Chen P, Noble RJ. A solid state accelerator. AIP Conf. Proc. 1987;**156**: 222. DOI: 10.1063/1.36458

[44] Östling D, Tománek RA. Electronic structure of single-wall, multiwall and filled carbon nanotubes. Physical Review B. 1997;**55**(20):13980-13988. DOI: 10.1103/PhysRevB.55.13980

[45] Tajima T, Cavenago M. Crystal X-ray accelerator. Physical Review Letters. 1987;**59**(13):1440-1443. DOI: 10.1103/PhysRevLett.59.1440

[46] Chen P, Noble RJ. Crystal channel collider: Ultra-high energy and luminosity in the next century. AIP Conf. Proc. 1997; **398**:273. DOI: 10.1063/1.53055

[47] Borrmann G. Die Absortion von Röntgenstrahlen im Fall der Interferenz. Zeitschrift für Physik. 1950;**127**:297-323. DOI: 10.1007/BF01329828

[48] Campbell HN. X-ray absorption in a crystal set at the Bragg angle. Journal of Applied Physics. 1951;**22**:1139. DOI: 10.1063/1.1700122

[49] Strickland D, Mourou G. Compression of amplified chirped optical pulses. Optics Communications. 1985;**56**(3):219-221. DOI: 10.1016/0030-4018(85)90120-8

[50] Mourou G, Mironov S, Khazanov E, Sergeev A. Single cycle thin film compressor opening the door to Zeptosecond-Exawatt physics. The European Physical Journal Special Topics. 2014;**223**:1181-1188. DOI: 10.1140/epjst/e2014-02171-5

[51] Li P et al. Recent advances in focused ion beam nanofabrication for nanostructures and devices:

Fundamentals and applications. Nanoscale. 2021;**13**:1529-1565. DOI: 10.1039/D0NR07539F

[52] Pint CL et al. Dry contact transfer printing of aligned carbon nanotube patterns and characterization of their optical properties for diameter distribution and alignment. ACS Nano. 2013;**4**(2):1131-1145. DOI: 10.1021/nn9013356

[53] Chen X et al. Atomic layer lithography of wafer-scale nanogap arrays for extreme confinement of electromagnetic waves. Nature Communications. 2013;**4**:2361. DOI: 10.1038/ncomms3361

[54] Pizzi A et al. Graphene metamaterials for intense, Tunable, and compact extreme ultraviolet and X-ray sources. Advancement of Science. 2020;**7**:1901609. DOI: 10.1002/advs.201901609

[55] Kong L-B, Chen Z. Plasmonic electron acceleration with the meta-surfaces. Physics of Plasmas. 2017;**24**:083111. DOI: 10.1063/1.4997481

[56] Bontoiu C et al. TV/M Laser-Driven Accelerating Gradients in Graphene. Proceedings of the 10th International Particle Accelerator Conference (IPAC'22); 12-17 June 2022; Bangkok, Thailand: Paper WEPOST043

[57] Sarid D, Challener W. Modern Introduction to Surface Plasmons: Theory, Mathematica Modeling, and Applications. Cambridge: Cambridge University Press; 2010. 371 p. DOI: 10.1017/CBO9781139194846

[58] Ukhtary MS, Saito R. Surface plasmon in graphene and carbon nanotubes. Carbon. 2020;**167**:455. DOI: 10.1016/j.carbon.2020.05.019

[59] Macchi A. Surface plasmons in superintense laser-solid interactions. Physics of Plasmas. 2018;**25**:031906. DOI: 10.1063/1.5013321

[60] Nejati M et al. The single wall carbon nanotube waveguides and excitations of their $\sigma + \pi$ plasmons by an electron beam. Physics of Plasmas. 2009;**16**: 022108. DOI: 10.1063/1.3077306

[61] Fedeli L et al. Electron acceleration by relativistic surface Plasmons in laser-grating interactions. Physical Review Letters. 2016;**116**:015001. DOI: 10.1103/PhysRevLett.116.015001

[62] Yakimenko V et al. FACET-II facility for advanced accelerator experimental tests. Physical Review Accelerators and Beams. 2019;**22**:101301. DOI: 10.1103/PhysRevAccelBeams.22.101301

[63] Rider AE, Ostrikov K, Furman SA. Plasmas meet plasmonics: Everything old is new again. European Physical Journal D: Atomic, Molecular, Optical and Plasma Physics. 2012;**66**:226. DOI: 10.1140/epjd/e2012-30273-3

[64] Que W. Theory of plasmons in carbon nanotube bundles. Journal of Physics: Condensed Matter. 2002;**14**: 5239-5253. DOI: 10.1088/0953-8984/14/20/319

[65] Lehe R, Kirchen M, Andriyash IA, Godfrey BB, Vay J-L. A spectral quasi-cylindrical and dispersion free particle-In-cell algorithm. Computer Communications. 2016;**203**:66-82. DOI: 10.1016/j.cpc.2016.02.00

[66] Sahai AA, Tajima T, Taborek P, Shiltsev VD. Solid-state tube wakefield accelerator using surface waves in crystals. International Journal of Modern Physics A. 2020;**34**:1943009-1943029. DOI: 10.1142/9789811217135-0009

[67] Sahai AA. Nanomaterials based Nanoplasmonic accelerators and light-sources driven by particle-beams. IEEE Access. 2021;**9**:54831-54839. DOI: 10.1109/ACCESS.2021.3070798

[68] Resta-Lopez J et al. Study of Ultra-High Gradient Acceleration in Carbon Nanotube Arrays. Proceedings of the 9th International Particle Accelerator Conference (IPAC'18); 29 April-4 May 2018; Vancouver, Canada: Paper TUXGBE2. DOI: 10.18429/JACoW-IPAC2018-TUXGBE2

[69] Zhang X et al. Particle-in-cell simulation of X-ray wakefield acceleration and betatron radiation in nanotubes. Physical Review Accelerators and Beams. 2016;**19**:101004. DOI: 10.1103/PhysRevAccelBeams.19.101004

[70] Roth M, Schollmeier M. Ion Acceleration–Target Normal Sheath Acceleration. Proceedings of the CAS-CERN Accelerator School: Plasma Wake Acceleration; 23-29 November; Geneva, Switzerland. DOI: 10.5170/CERN-2016-001.231

[71] Nicks BS et al. High-density dynamics of laser wakefield acceleration from gas plasmas to nanotubes. Photonics. 2021;**8**:216. DOI: 10.3390/photonics8060216

[72] Shin YM. Optically controlled coherent X-ray radiations from photo-excited nanotubes. Nuclear Instruments and Methods in Physics Research Section B: Beam Interactions with Materials and Atoms. 2017;**407**:276-281. DOI: 10.1016/j.nimb.2017.07.019

Chapter 5

Faster Fusion Power from Spherical Tokamaks with High-Temperature Superconductors

Colin Windsor and Gurdeep Kamal

Abstract

The use of spherical tokamaks and high-temperature superconductors (HTSs) offers the possibility of achieving faster fusion power by allowing plants of high-field, high plasma pressure, and good energy confinement thereby reducing the need for large plasma volumes. This spatially efficient energy-dense approach accesses quicker development and the possibility of modular construction. An overview of high-performance computational (HPC) capabilities at Tokamak Energy is given. We describe, at a highlevel and in practical terms, the use of theory, models, algorithms, and applications to develop spherical tokamak designs in an integrated fashion. A challenge of spherical tokamaks is that there is less room for the neutron and gamma shield necessary to prevent heating and radiation damage to the HTS core. Tungsten boride shield materials may be able to provide an optimal combination of inelastic (n, gamma) reactions and gamma attenuation. The neutron energy is reduced largely by inelastic reactions to energies where boron absorption occurs, while tungsten attenuates the resulting gammas rapidly. Although inelastic scattering is shown to be the key to tungsten boride shield performance, it is shown that the remaining neutrons generated in the plasma and transmitted without reaction through the shield dominate the heat deposition in the HTS core.

Keywords: fusion, spherical tokamaks, tungsten borides, shielding, neutrons, gammas

1. Introduction

Fusion power is one of the few options capable of supplying abundant safe baseline energy to the world [1]. In 1955, John Lawson, working at Harwell, defined a condition for fusion power depending on plasma density n and temperature T. It was later extended to include confinement time τ_E., to form what is now known as the "triple product" $nT\tau_E$ which must exceed 3.10^{21} keV m^{-3} s for fusion reactions to be self-sustaining [2]. Costley, Hugill, and Buxton [3] showed that this triple product is proportional to the equivalent triple product $\beta B_T^2\tau_E$ where β is the ratio of the plasma pressure, $p_{pla} = n.kT$, to the magnetic pressure $p_{mag} = B^2/ 2\,\mu_0$. B_T is the toroidal field and τ_E is the energy confinement time, related to the plasma volume, kT is the thermal energy, and μ_0 is the vacuum permeability constant. They showed that the high

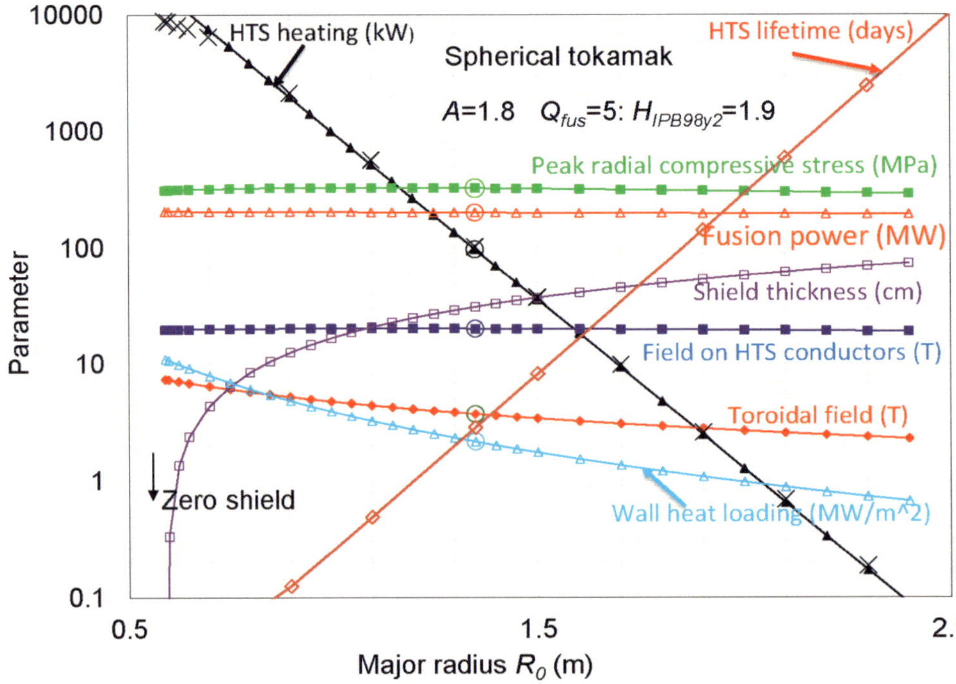

Figure 1.
Some example results from the 2017 Costley system code scanning as a function of major radius for a constant fusion gain Q_{fus} = 5, *spherical tokamak aspect ratio* A = 1.8, *and confinement scaling* H_{IPB9y2} = 1.9. *In this series, the core diameter is increased by 8% for each change in major radius to keep the peak radial compressive stress (green squares) approximately constant. The fusion power* P_{fus} *(red open triangles) is seen to be roughly independent of major radius. The shield thickness (violet open squares) increases, initial rapidly from zero, with major radius. Also shown are the wall heat loading (sky blue open triangles) and the toroidal field* B_T *(blue closed squares). Incorporated into the code are the HTS core heating (black closed triangles, with individual calculations as crosses) and the estimated HTS lifetime from the fast neutron fluence (pink open diamonds).*

β values possible with spherical tokamaks and the higher toroidal fields B_T possible with high-temperature superconductors (HTSs) were a viable alternative to ever-increasing tokamak size in the search for fusion power.

Their method in [3] was to write a "system code," in their case a simple spreadsheet, to incorporate the key input parameters and equations to calculate key output parameters. Innovations were to use the fusion gain Q_{fus} as an input parameter to their calculations and to introduce checks to ensure that the actual plasma density was never above 0.8 of the known Greenwald density limit and the plasma normalized pressure β value never above 0.9 of the Troyon β limit to plasma stability. Since high density and high plasma pressure are both desirable for high fusion gain, the best plasma conditions would be those that would approach these limits.

Some example results using their code are shown in **Figure 1**. The fusion power with these scan conditions is seen to be almost independent of tokamak size, while key engineering conditions, such as the radial compressive stress at the top of tokamak, the magnetic field at the HTS conductors, and the wall loading power are all satisfactory. The purple open squares show the thickness of the shield, rising rapidly from 0.55 m major radius.

2. The use of modeling codes to optimize spherical tokamak designs

Figure 1 describes how an integration of tokamak physics, HTS design, neutronics and computational fluid dynamics, and magnetohydrodynamic stress codes could help predict optimal tokamak design. To these, we must add codes for tritium breeding, heat transfer, design, construction, and decommissioning costs to produce a suite of codes that must be put together to predict a new spherical tokamak power plant design. It is with this aim that Tokamak Energy has invested in high-performance computational (HPC) capabilities.

The "system codes" which used to be implemented on Excel spreadsheets are greatly expanded and are verified by fitting experimental data from actual tokamaks like ST40. Future system codes are envisaged to use a single-design geometry to perform calculations automatically on a suite of tools involving physics, HTS magnet design, neutronics, heat flow, and engineering to obtain the parameters required to optimize a tokamak. The integration of system codes and tools used by multiple departments allows for an automated, iterative optimization of the tokamak design. HTS magnet codes need to consider quench protection and be compared to actual demo magnet experiments to verify their performance. The analytic stress equations, once commonly used, are now replaced by finite element codes whose performance can be judged by experimental data from the ST40 tokamak. The neutronics of HTS and biological shields is now rather straightforwardly assessed by activation foils under actual conditions and actual measurements of dose rates during shots. A major part of the successful operation of tokamak power plants depends on being able to breed tritium from the fusion neutrons interacting with lithium in a "blanket" within the biological shield. These are being designed using neutronics codes but to date have not been tested. "Process" codes have the particularly difficult job of computing the "cost of electricity" for given costs of capital borrowing and materials. They are not easy to check and can so easily be outdated by unforeseen world events. The integration of system codes and tools used by multiple departments allows for an automated, iterative, optimization of the tokamak design.

All these aspects of physics, engineering, heat flow, magnet design, neutronics, radiation damage, shielding, tritium breeding, and cost estimates are possible with separate codes, and the objective is to integrate them into a suite of codes using the same computer-aided design (CAD) geometry on the same hardware. For this purpose, a high-performance computer was purchased by Tokamak Energy. The Tokamak Energy HPC currently has over 600 cores and over 3 TB of RAM available to use and continuously uses thousands of CPU hours daily to resolve physics and neutronics problems. **Figure 2** illustrates some of the principal applications and their codes used, with illustrations of the results they have given in hours by the high-performance computer illustrated at the center.

Key technological advancements needed to complete integration of CAD geometry between the suite of codes is in mesh generators. Currently, all tools have their own meshing methods with outputs in various formats. Geometrical information required to build successful inputs to the system tools are envisaged to be integrated into the CAD data. For example, neutronics heat deposition calculations require material compositions and densities for input and can output maximum heat deposition per CAD part which can then be stored as part of the CAD data for input for heat flow tools. Upgrades to the system code will utilize this data and perform meshing to the required resolution for each CAD part which are required for the system tools.

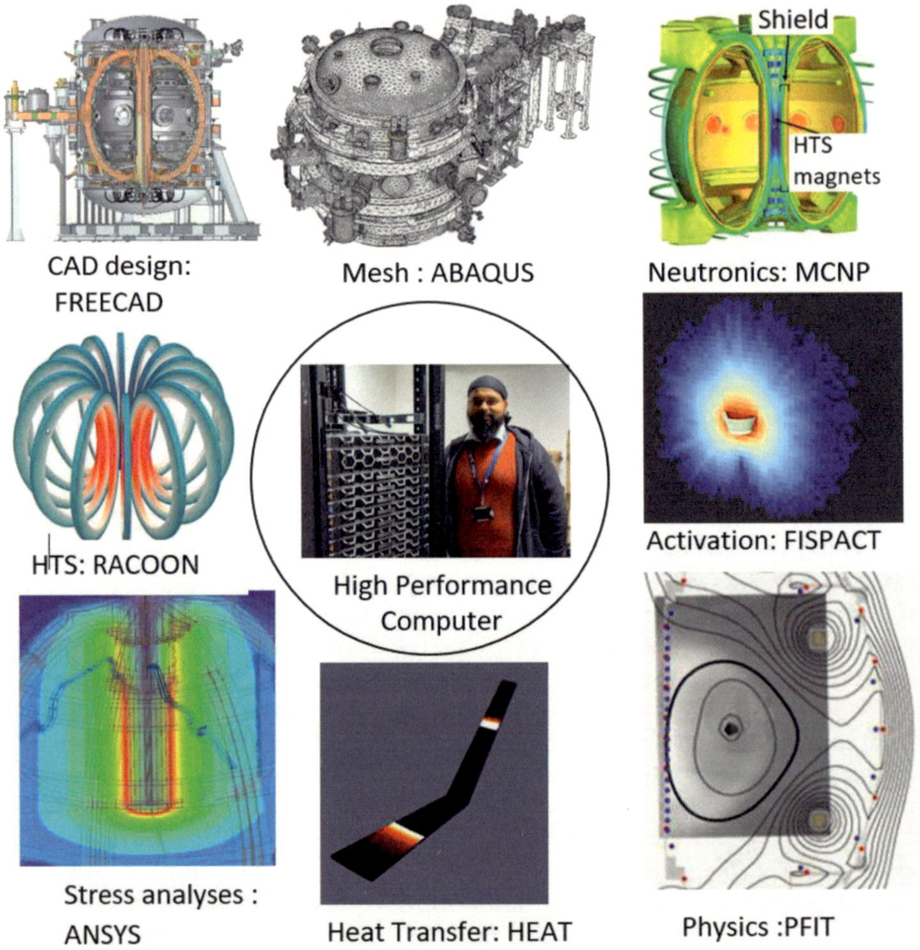

Figure 2.
The Tokamak Energy high-performance computer and examples of some of its applications and widely used code. Clockwise from the top left; the design "FREECAD" computer-aided design code of ST40, an "ABAQUS "code image of the mesh representation of the design, "MCNP" the Monte Carlo Neutral Particle code showing the neutronics fluence of a possible power plant, "FISPACT" the fusion activation code plot of the ST40 inner vacuum vessel, the in-house "PFIT" code plots of the ST40 field line fits to ST40 image, heat transfer in the ST40 divertor, the "HEAT" heat transfer code plots of the temperature in the HTS coils of a power plant, the "ANSYS" code plot of the mechanical stresses in the ST40 tokamak, and the "RACOON" in-house code for calculating HTS performance. In the center is the Tokamak Energy high-performance computer with author GK.

The following sections will shed some light on how we are effectively using the HPC for streamlining the design process, with an emphasis here on shield design. The use of "tally tagging" with the neutronics code MCNP is to answer key questions on the performance of shields, such as where the neutrons and gammas responsible for core heating originate, and from what reactions.

3. Spherical tokamaks

Costley [4] showed that the triple product $nT\tau_E$ could be expressed in terms of the H factor, the safety factor q, the major radius R_0, the toroidal field B_T, aspect ratio A,

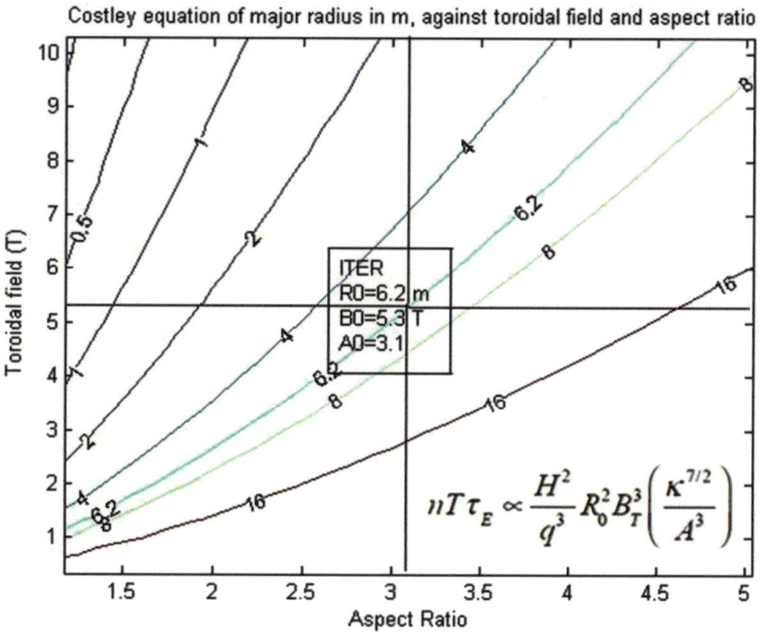

Figure 3.
The reduction in tokamak major radius (lines labeled in meters) which may be given by lower aspect ratios (X-axis) or by higher magnetic toroidal fields (Y-axis). The figure uses the equation shown from [4]. The data are scaled to the ITER operating point at major radius R_0 = 6.2 m, toroidal field B_0 = 5.3 T, and aspect ratio A = 3.1.

and the plasma elongation κ as $nT\tau_E = H^2.q^{-3}R_0^{\ 2}B_T^{\ 3}\kappa^{7/2}A^{-3}$. It is therefore possible to explore the dependence of the tokamak major radius R_0 as function of the tokamak aspect ratio and toroidal field at a constant value of the triple product, safety factor, and H factor. For example, those assumed for ITER are shown in **Figure 3**.

It is seen from **Figure 3** that going to lower aspect ratios (leftward) or to higher toroidal fields (upward) gives similar triple products from smaller tokamaks with lower major radii. This result is important because tokamaks of ITER size may be defined as "megaprojects" (technically project over $1B: ITER is over $20B). Historical studies of megaprojects show that cost and time overruns are the norm rather than the exception [5]. The reasons are obvious enough: megaprojects tend to be too big for one company, or even one country. They become difficult to manage as the layers of management between boss and worker become longer. With smaller tokamak power plants, agility becomes possible when projects can adapt quickly to any design optimization. A breakthrough would be achieved when modular construction becomes possible with centralized manufacture remote from the construction site.

In the following sections, these two opportunities, spherical tokamaks and high-temperature superconductor magnets, will be explored in more detail. **Figure 1** does also reveal the downside of spherical tokamaks. The purple open squares in **Figure 1** show that the space available for shielding diminishes slowly at large major radii but eventually rather rapidly to zero at some smaller major radius. Much of the remainder of this chapter will explore the materials and mechanisms most appropriate to a fusion tokamak shield. Tungsten boride shields have proved one possible option for a fusion power plant, and their performance is examined and compared with other possibilities.

In 1983, Sykes, Turner and Patel [6] and later Peng and Strickler [7] predicted the high plasma pressures β possible with low aspect ratio spherical tokamaks. The theory showed that β could indeed be raised by a combination of low aspect ratio and high elongation. In 1998, the START spherical tokamak team at Culham Laboratory in the UK achieved a breakthrough by reaching β values of order 40% [8], compared to the 6% or so achieved in JET or the 4% or so expected in ITER.

A simple-minded explanation for these increases in efficiency is suggested by **Figure 4** which shows the toroidal magnetic field B_T in Tesla around a center column carrying a constant current I as a function of radius r. The field is given by Ampere's law $B = 2\,\mu_0.I/4\pi r^2$ or $B\ (Tesla) = 0.2.I\ (MA)/\mathrm{r}\ (\mathrm{m})^2$, so the smaller the radial distance the higher the field for a given center column current. For a current of 10 MA, the field is 2 Tesla at a radius of 1 m. A plasma with higher elongation κ and triangularity ε naturally squashes the plasma so that the orbiting electrons spend more time close the center column and make a more efficient use of the available field. Another advantage allowing the prospect of a power plant with reduced external current drive was the experimental observation of the bootstrap current, driven by plasma pressure gradients, which had been predicted theoretically by Bickerton, Connor, and Taylor in 1971 [9] and observed experimentally on the Princeton tokamak TFTR by Zarnstorff et al. [10] in 1988.

Bigger spherical tokamaks were soon built around the world, now upgraded to higher toroidal fields. These include MAST-UPRADE (0.55 Tesla) at Culham, UK [11], NSTX-U (1Tesla) at Princeton, USA [12], and GLOBUS M2 (0.82 Tesla) at St. Petersberg, Russia [13]. Most recently constructed with the highest design toroidal field of 3 Tesla is the ST40 at Tokamak Energy Ltd., Milton Park, UK. This company was started in 2009 and

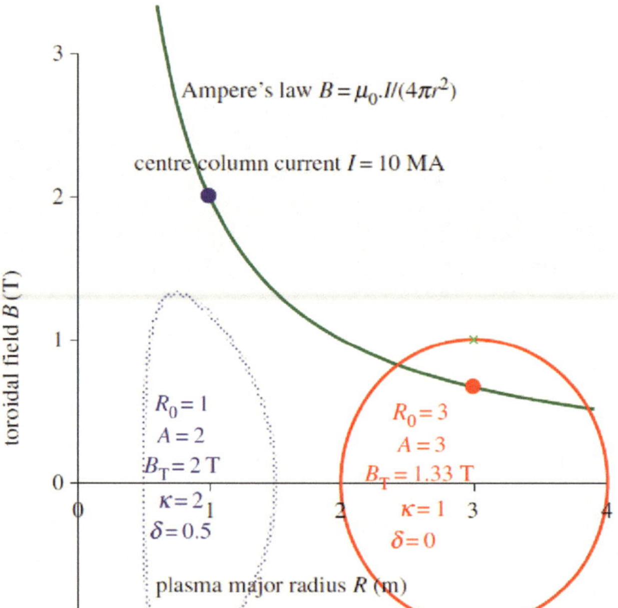

Figure 4.
In a low aspect ratio, high elongation plasma (left in blue), the orbiting electrons spend more time close to the current carrying center column than in a conventional tokamak (right in red) from [1]. Since the toroidal magnetic field drops off inversely with radius, this means that a given center column current gives a bigger field.

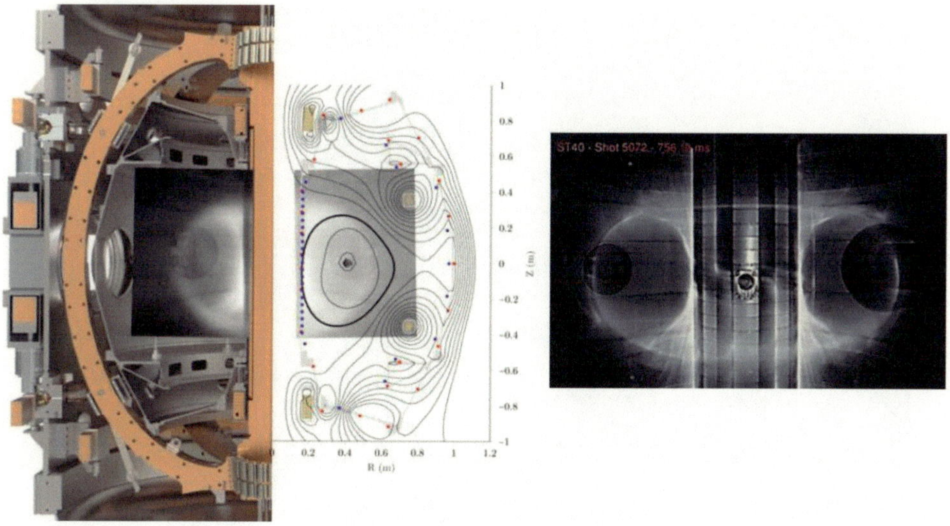

Figure 5.
The ST40 tokamak in May 2019 when the milestone plasma temperature of 15x10^6 K was achieved. On the left is the engineering drawing along with the reconstruction of the plasma and optical image of the plasma.

its founders included START pioneers Alan Sykes and Mikhail Gryaznevich [14]. Its aim was to exploit the twin technologies of spherical tokamaks and high-temperature superconductors. The ST40 design team was chaired by Alan Sykes in 2014. Engineer and draftsman John Ross rapidly iterated designs with pencil and drawing board which hardly changed when the design moved to computer-aided design. **Figure 5** shows the design as of May 2019 when temperatures of 15x10^6 K were first achieved.

Rapid development followed and ST40 moved to larger premises with space for a neutron shield and neutral beam injection to heat the plasma to the next milestone temperature of 100x10^6 K which was successfully achieved in February 2022. In designing the next generation of spherical tokamaks, HPC facilities will be important particularly in solving the gyro-kinetics used to predict the plasma turbulence and plasma confinement time.

4. High-temperature superconductors

Bednorz and Müller discovered high-temperature superconductors (HTSs) back in 1986 [15]. The theory remains uncertain. The important fact is that by 2010 it was possible to buy kilometer lengths of HTS tape just 0.1 mm thick and 12 mm wide. Under the optimized manufacturing conditions, now readily achieved by commercial suppliers, $YBa_2Cu_3O_7$ (YBCO) has a superconducting critical temperature of ~90 K. State-of-the-art commercial tapes are now exhibiting engineering critical current densities >1000 A/mm^2 at 20 K and 20 Tesla field applied perpendicular to the tape which is the lowest performance configuration [16].

In 2015, Tokamak Energy demonstrated a simple 25 cm radius HTS tokamak operating continuously for 29 hours under a live demonstration during the Royal Society Summer Exhibition as illustrated in **Figure 6**.

The more difficult task was to create the HTS coils needed for an operational tokamak. A key problem is quench protection when any defect or heating anomaly

Figure 6.
The high-temperature superconducting tokamak ST25HTS in operation after for 29 hours.

can create a local temperature rise above the transition temperature. With an insulated conductor, the current has nowhere to go and so it heats up quickly. This can spread to nearby superconductors and so the magnet quenches and could be destroyed. The problem can be avoided by removing any insulation between the tapes. The current from any localized normal region of tape can then spread over to neighboring tapes and any

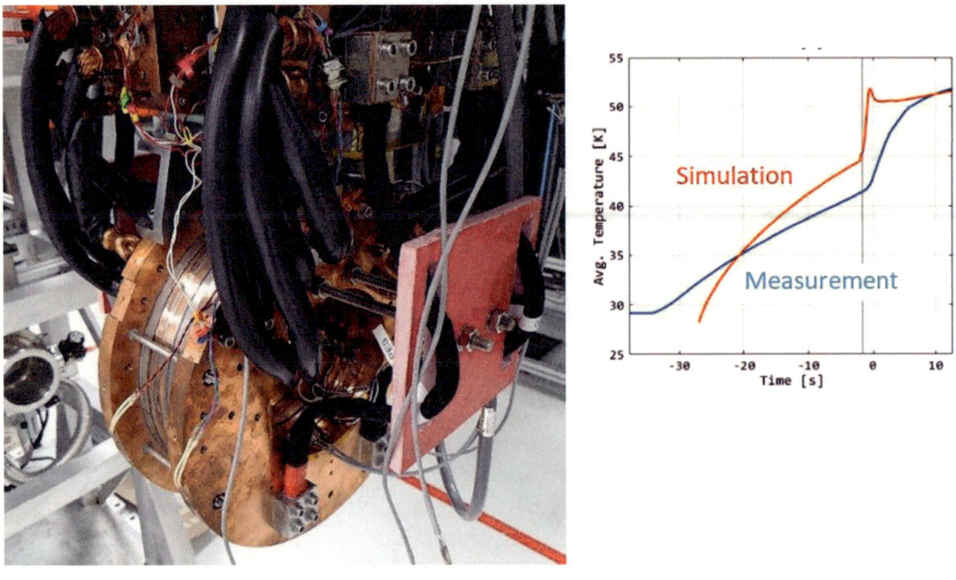

Figure 7.
The DEMO2 magnet constructed at Tokamak Energy [17]. The coil has a bore of 300 mm and contains six modular "pancake" coils wound with 738-m of 12-mm-wide tapes with variable inter-turn resistance between tapes and reached 10 Tesla. Inset is shown the RACOON code predictions of the coil temperature compared with the measurements during a quench at the vertical line position.

quench is avoided, but then the time constant for energizing the coil becomes impractically long. In research in the David Hawksworth HTS Magnet Laboratory at Tokamak Energy, a middle way of partial insulation has been developed with a precise inter-turn resistance between each tape layer to allow current transfer but avoid the time-constant problems. **Figure 7** shows "Demo2", a stack of six REBCO non-insulated pancakes with a diameter of 300 mm. Cooled to 20 K it is able to take a current of 3000 A and produce a peak field of over 10 Tesla. The magnet proved difficult to quench. It could be quenched by turning off the current, but the stored energy was dissipated within the magnet in the inter-turn resistance, and there was no system degradation [18].

Under construction at Tokamak Energy is a demonstration tokamak "Demo4" designed to operate at fields in the region of 20 Tesla with a full set of partially insulated HTS toroidal field coils.

5. Shield design explored using "tally tagging"

Figure 8 illustrates a well-known challenge of spherical tokamaks. For a given central HTS core radius R_{core}, (blue) and necessary vacuum *gap* (white) between the plasma (yellow) and the outer shield surface (brown), the available shield thickness drops rapidly with aspect ratio A. The HTS coils need to be actively cooled to around 20 K to remove the heat deposited from the incoming neutron and gamma irradiation. Also, the performance of the HTS material can be degraded by radiation.

A related dependence was shown by the open purple squares in **Figure 1** as a function of major radius for constant aspect ratio $A = 1.8$. **Figure 1** illustrates the balance that tokamak design must take between the HTS design constraints limiting the fields possible with a given core radius, the physics constraints of fusion power, wall heat load, the neutronics constraints of heat deposition into the HTS core, radiation damage limitation, and limiting the engineering stresses to acceptable levels. Most of these improve with larger major radius, but not all. The stored energy of the plasma increases as the cube of the size, but the surface area as the square, so that the energy

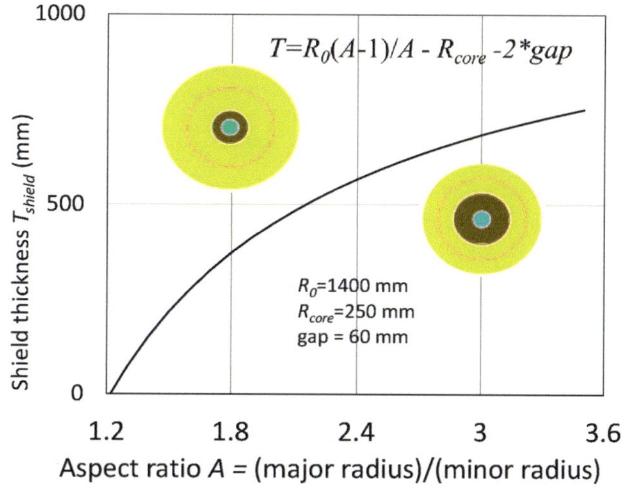

Figure 8.
The increase of shield thickness T with aspect ratio A for tokamaks of fixed major radius R_0 = 1400 mm (red line) and fixed HTS core radius R_{core} = 250 mm (blue).

per surface area increases with size [19]. This makes plasma disruptions, where some instability makes the plasma touch the wall and release its energy, a serious problem with large tokamaks and less serious with more compact tokamaks.

The main disadvantages of large size are cost and construction time, which did not appear in **Figure 1**. In practice, deciding the optimal size of a spherical tokamak power plant will be an optimization including these as well as outputs from plasma physics, HTS core construction, neutronics calculations, heat flow calculations, and structural engineering constraints.

The fusion power plants envisaged to use the D-T reaction when a deuterium and tritium ions collide to produce an alpha particle and a neutron $^{2}D + ^{3}T = ^{4}He + n + 17.6$ MeV. The helium alphas, being charged, tend to stay within the plasma and sustain its temperature. The neutrons, with 14.1 MeV energy, interact little with the plasma and escape to the inner and outer shields where their energy can be used to generate electricity and also to replenish the tritium ions by reactions with lithium in "blanket" layers within the shields. There are fewer size limitations to the outer shield, so this discussion will center on the optimization of the inner shield.

14.1 MeV neutrons are not easy to shield. They can be slowed down by multiple elastic collisions, moderation, or more quickly by inelastic (e.g. n, gamma) collisions. The moderation method is only practical for hydrogen moderators as the average value of the decrease in the natural logarithm of the neutron energy per collision is unity for natural hydrogen, but only 0.158 for carbon. The shield studies made for the Costley system code [3] results of **Figure 1** supposed a layered shield with thick layers of cemented tungsten carbide and thinner layers of water. The fast neutron flux through such a shield is shown in **Figure 9** taken from [20]. The half-attenuation

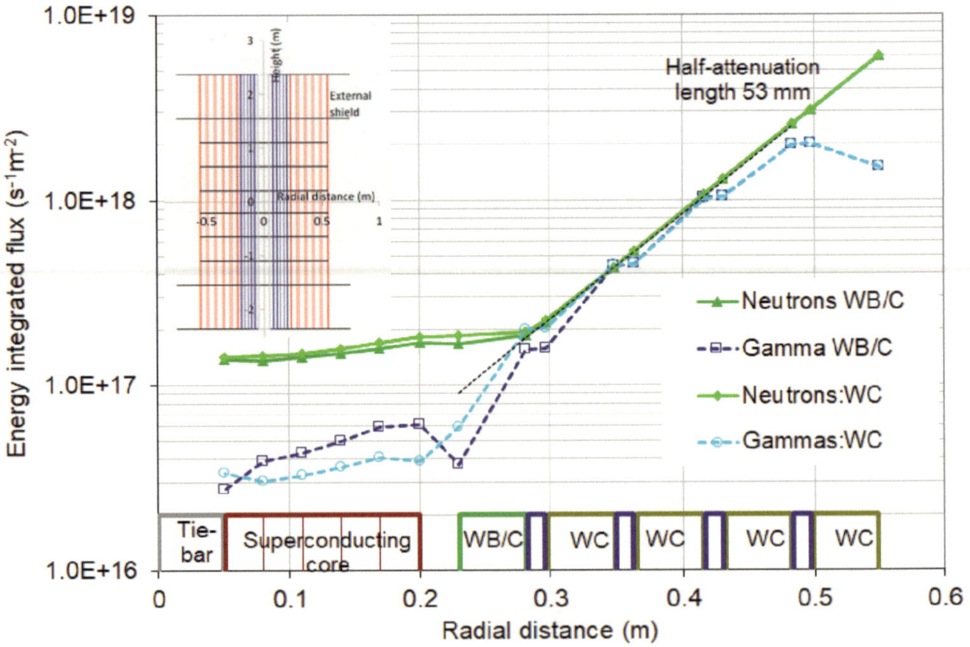

Figure 9.
The attenuation of fast neutrons (E > 0.1 MeV) though a layered shield of tungsten carbide (including with a tungsten boride inner layer) with intervening water channels for coolant and neutron moderation. The inset shows a vertical section [20].

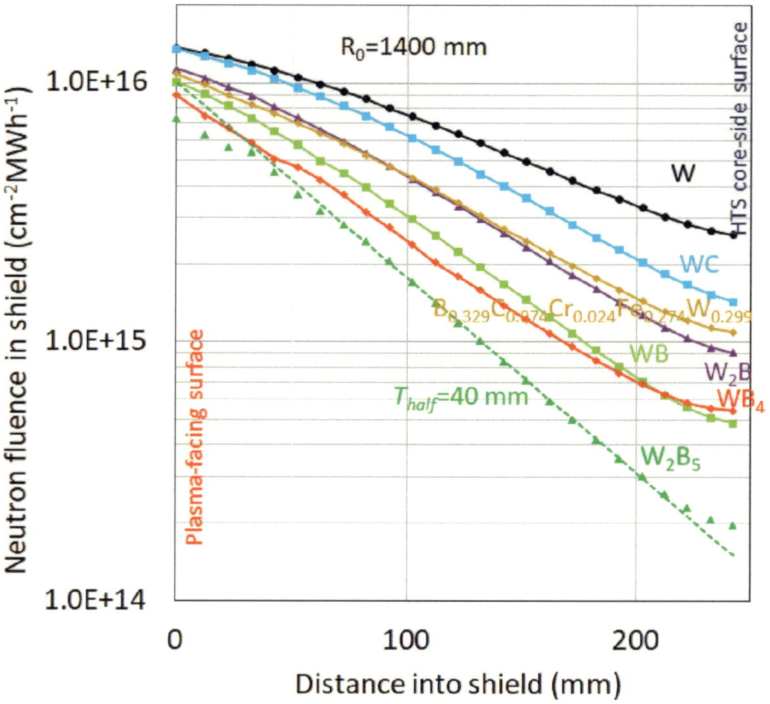

Figure 10.
The attenuation of neutrons though monolithic shields at 1400 mm major radii of boron carbides of varying composition, including a cemented borocarbide, pure tungsten, and tungsten carbide.

shield distance is 53 mm. This relatively poor attenuation is possibly because of the relatively high atomic fractions of carbon and oxygen which give little attenuation to either neutrons or gammas.

Water coolant is now seen to have serious problems in a fusion power plant. The activation of ^{16}O to ^{16}N which decays with a 7.1-s half-life is one issue, and another is the desire for heat production at higher temperatures where the water vapor pressure becomes serious. Helium gas cooling has none of these problems.

Monolithic shields with varying fractions of tungsten and boron were examined [21, 22] and shown to give half-attenuation coefficients as low as 40 mm as seen in **Figure 10**. The neutron cross sections of tungsten and boron are shown as a function of energy in **Figure 11**. Tungsten is a key element as its neutron cross section has significant inelastic neutron (n, 2n) and (n, n' gamma) reactions around the 1 to 10 MeV energies which reduce the 14.1 MeV neutrons to around 0.1 MeV energy. The ^{10}B isotope has an absorption cross section which rises inversely with neutron velocity giving a highly significant absorption below 0.1 MeV, not far above the energies where they are captured by ^{10}B which makes up around 20% of natural boron. The principal reactions occurring in a tungsten boride shield are illustrated in **Figure 12**. Let us consider each of the numbered processes.

Elastic scattering (i) of neutrons. Tungsten makes a good reflector as its cross section of ~3 barns of elastic cross section per atom at 14 MeV and its high mass compared to a neutron ensures little recoil energy. Most neutrons are not reflected but continue forward with slightly reduced energy. Moderation (ii) occurs when incident neutrons collide with lighter atoms, like boron. The collisions exchange, or moderate, the neutron energy, therby reducing the transmitted power. The best moderator is

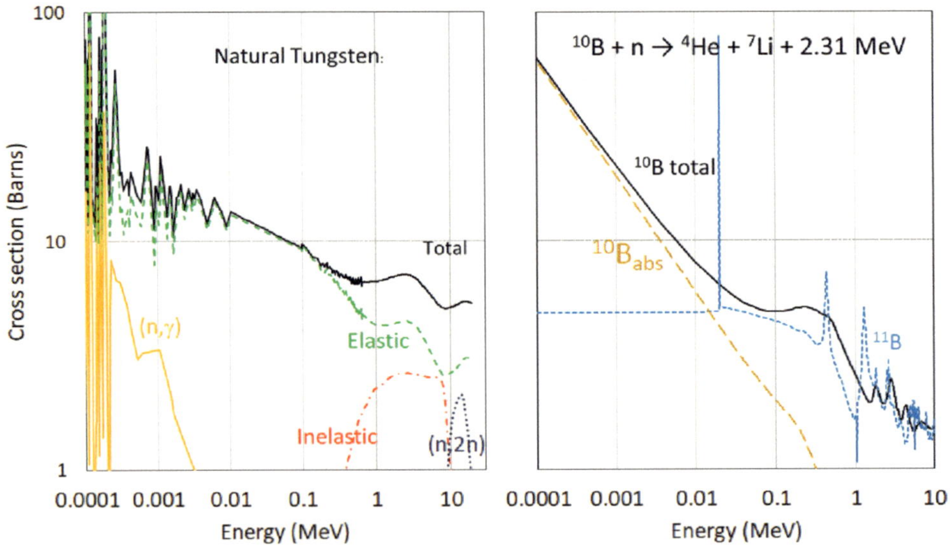

Figure 11.
The neutron cross sections for tungsten (left) and for boron (right). For tungsten, the total (black), elastic (green, dashed), inelastic (red, dash-dot), (n,2n) (blue, dotted), and (n, gamma) (yellow) cross sections are shown. For boron, the minority (20%) ^{10}B isotope (black) and the majority ^{11}B isotope cross section (blue dashed) are shown. Below 10 keV, the ^{10}B cross section follows an inverse velocity increase with decreasing neutron energy as show in the curve labeled $^{10}B_{abs}$. The cross sections are from the Brookhaven National Nuclear Data Centre [23].

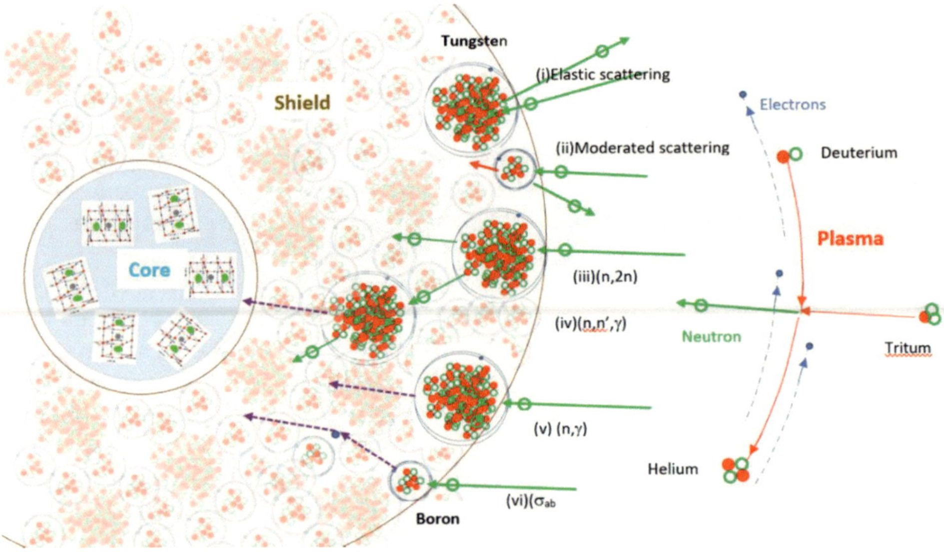

Figure 12.
Some of the processes involved in a tungsten boride shield protecting a high-temperature superconducting core from fusion neutrons generated by a plasma. Tungsten atoms are large, while boron atoms are small. Protons are red, neutrons green, electrons blue, and gammas violet. Fusion is fueled continuously by injecting tritium ions usually by pellet injection. These hit deuterium ions to produce helium ions and fast neutrons. The six scattering processes indicated are discussed in the text.

hydrogen since its mass is almost the same as the neutron and the maximum energy is lost on collision. Inelastic (n,2n) (iii) reactions produce a different isotope of the same element. A typical reaction would be $^{183}W + n = {}^{182}W + 2n + \gamma$, which is highly beneficial transforming each high energy neutron into two lower-energy neutrons of a few MeV which are easier to shield.

Inelastic (n, n' gamma) neutron scattering (iv) means that the incident neutron forms a new compound nucleus which quickly decays to release the neutron with an appreciably lower energy and with the emission of the excess energy in the form of a gamma. These gammas must then be shielded. (n, gamma) capture (v) is common at lower neutron energies. The neutron is absorbed to form a tungsten isotope with one higher atomic weight with the emission of gammas of significant energy (vi). The lower-energy neutrons produced by moderation or inelastic scattering become increasingly likely to be absorbed by isotopes such as ^{10}B with a high neutron absorption cross section. Gammas are shielded by scattering from electrons in the shield atoms and thus need high atomic number (number of electrons per atom) and high number density (atoms/m^3). Tungsten is a comparatively good element for a gamma shield, while boron is not.

More detailed information on the shield scattering process may be found by using the "tally tagging" feature of the MCNP code [24]. Fluence or energy deposition tallies generally total up all particles from all sources, reactions, atoms, and isotopes. Using tally tagging, the results can be broken down into any chosen set of these options. In order to limit the output file length, it is usually best to ask specific questions: "Did a neutron start in the plasma and survive all the way through the shield?" "Did a photon heating the HTS core get born in the core itself, or did it get born in the shield?" Each output contains a tally code of form CCCCCZZAAA.RRRRR where CCCCC optionally represents the origin tally cell number, ZZ the atom number, AAA the isotope number, and RRRRR the reaction type.

Another important refinement in fluence tallies is to separate them according to the angle the particle path makes with the normal to the tally surface. In the following examples, the angles were divided into six ranges 0° to 30°, 30° to 60°, 60° to 90°, 90° to 120°, 120° to 150°, and 150° to 180°. These subtend different solid angles, so these need to be corrected to give fluences per steradian.

Some results answering the question of where the neutrons originate from are given in **Figure 13** which shows the fluences at angles $<30^0$ to the shield layers. The tally tagging distinguishes the tally volumes where each neutron originated. The red triangles show those neutrons originating in the plasma, although some elastic moderation will have occurred. The direct neutron energy spectrum at 100 mm into the shield shown inset in red at the bottom left of **Figure 13** shows that this is a minor effect with elastic moderation caused by the tungsten and boron atoms giving the rapid half-attenuations of 0.24 and 0.8 MeV, respectively. The corresponding indirect neutron spectrum shown inset in blue at the bottom left of **Figure 13** is from the tungsten (n, gamma) and (n, n' gamma) reactions and is very broad and centered around 0.2 MeV. The corresponding gamma rays are seen to have a less broad distribution with some peaks centered around 1 MeV. The tally tagging feature can also identify the origin of the indirect fluence in more detail. It shows that most of the indirect neutrons come back from tungsten atoms further into the shield. The indirect gammas came equally from tungsten and boron atoms within a few centimeters of the fluence position.

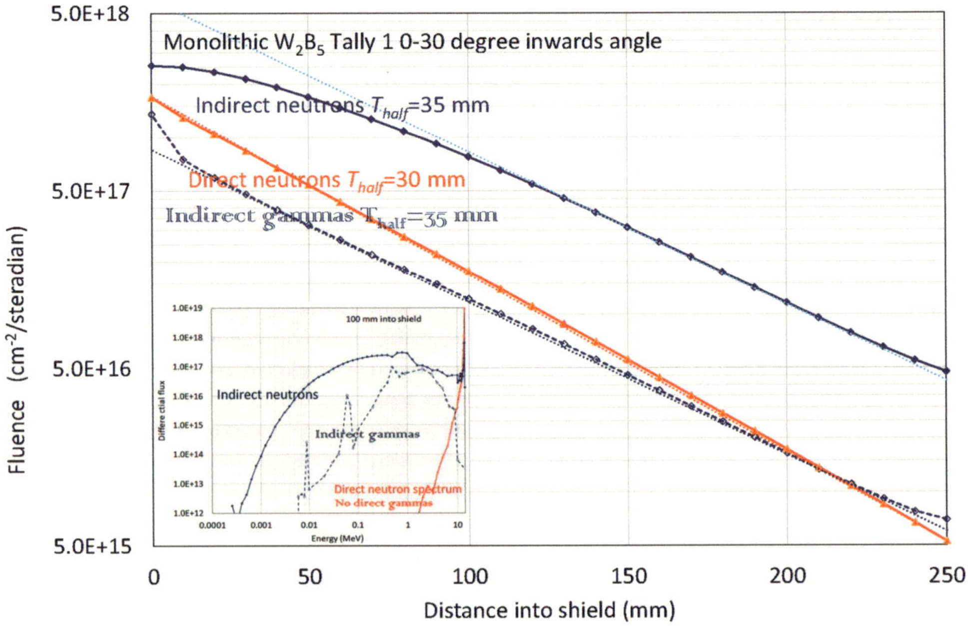

Figure 13.
The attenuation of fluence through a W_2B_5 shield for neutrons created in the plasma (red direct) and neutrons and gammas created inelastically in the shield (blue). There are no direct gammas. Inset are shown the energy spectrum of the direct neutrons (red upper right) and the indirect neutrons and gammas (blue lower left).

An illustration of the energy dependence of the neutrons as they pass through a W_2B_5 shield is given in **Figure 14**. Neutrons incident on the shield make up a "pool" of 14 MeV neutrons which extends with decreasing fluence through the shield. At any shield position, a fraction "falls down the waterfall" to energies around 0.2 MeV. At the base of the "waterfall," the neutrons lose further energy by moderation down a gently sloping riverbed. In W_2B_5 as their energy reaches 0.001 MeV energies, they soon become almost completely absorbed by ^{10}B as in a porous sandy riverbed.

A simple model of the W_2B_5 shield is that the fluence of direct neutrons of around 14 MeV energy exists across the shield with around 30-mm half-attenuation distance. Inelastic (n, n' gamma) reactions lower the neutron energy to around 0.2 MeV, but the resulting gamma rays are attenuated in only a few centimeters. Further moderation reduces the neutron energy to the level where they are absorbed by ^{10}B. In a pure tungsten shield, the waterfall from 14 to 0.2 MeV is again present, but there is no ^{10}B absorption, so once again low-energy neutrons build up steadily through the shield leading to an ever-increasing pool of low-energy neutrons.

W_2B_5 is exceptional in that its ^{10}B absorption removes most neutrons within a localized region of the shield so that the energy spectrum is almost identical through the bulk of the shield.

The HTS core of a tokamak power plant needs to be kept at temperatures of order 20 K. The heat deposition into the core determines the cryogenic power necessary to maintain this low temperature. The power deposition is also a determinant of the HTS radiation damage. The neutron and gamma power depositions into the core have been evaluated previously, but here the tally tagging option of MCNP has been used to investigate the processes by which this heat arrives at the HTS material. **Figure 15** shows the neutron and gamma contributions to the heat deposition into a fully cooled

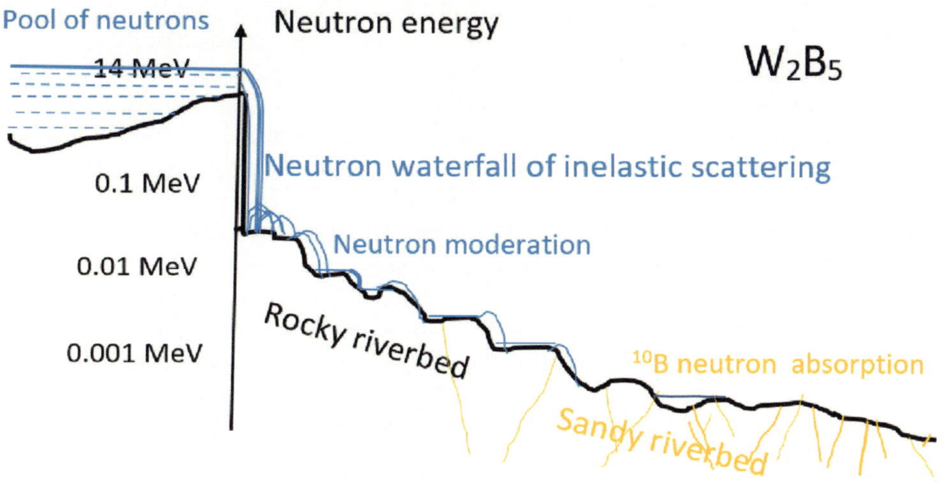

Figure 14.
An illustration of the energy dependence of the neutrons passing through A W_2B_5 shield. Neutrons incident on the shield come from a "pool" at 14 MeV. This pool decreases in width through the shield as (n, n' gamma) inelastic reactions produces an energy "waterfall" reducing the neutron energy to around 0.2 MeV. At the bottom of the waterfall, moderation reduces the energy gradually (the rocky riverbed) until energies are almost wholly absorbed by ^{10}B (the sandy riverbed). The gammas produced by the inelastic scattering are absorbed in a few centimeters so that the diagram is valid throughout the bulk of the shield.

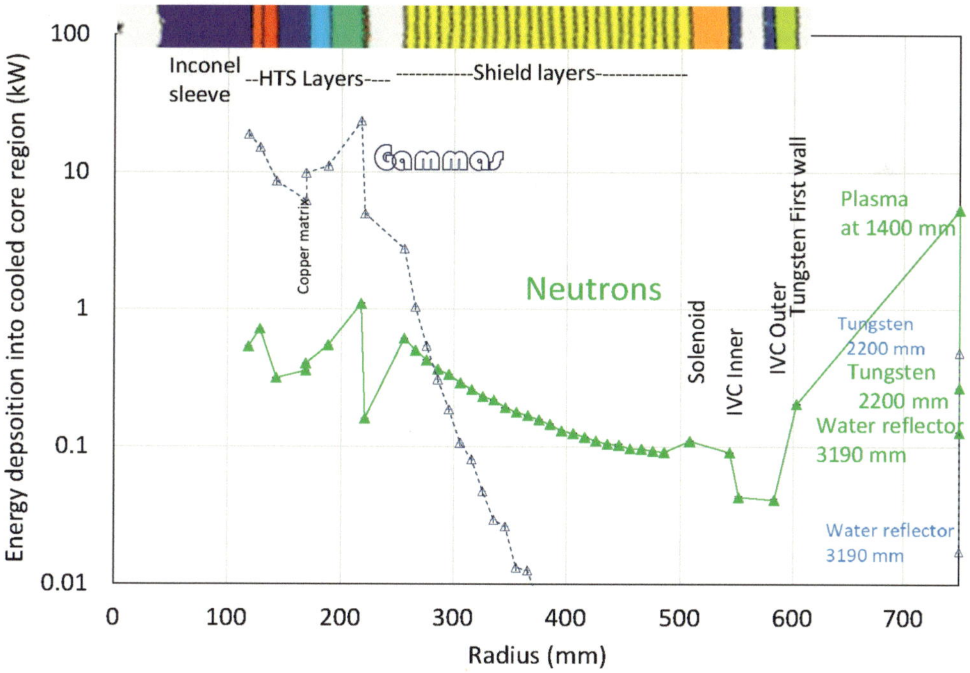

Figure 15.
The origins of the heat deposition into the cooled core of the tokamak including the inconel sleeve, HTS coils, copper cladding, and stainless inner vacuum vessel and shield layers as shown inset at the top of the figure. The major gamma component arises from the core components with the shield component decreasing rapidly. The neutron component is dominated by the direct contribution from the plasma itself and the outer wall reflector.

core region which includes the central inconel sleeve, the copper cladding surrounding the HTS, and the 316 L stainless inner vacuum vessel. It was established that the total HTS gamma heating was some four times larger than the total neutron heating [21]. It is clear from the figure that the largest gamma contributions arise around the HTS tape layers and the surrounding materials, copper cladding, and the inner vacuum vessel. Contributions from the shield layers are around an order of magnitude less and fall of rapidly into the shield. Direct gammas from the plasma and outer wall are very low.

In contrast, the major contribution to the neutron heat deposition is from the plasma itself out at 1400 mm radius (way off the scale of the figure!). As described earlier at each shield layer, some of these neutrons are inelastically scattered to give much lower-energy neutrons which are soon absorbed, and gammas which are rapidly attenuated within the shield. Some 14 MeV neutrons do persist all through the shield and provide the bulk of the heat deposition. Neutrons inelastically scattered by the HTS core volume, and the shield give contributions which are significant but around an order of magnitude lower.

Again, the tally tagging feature of MCNP can be used to explore in more detail the elements, isotopes, and reactions which give rise to the heat deposition. **Figure 16** shows the neutron (left) and gamma (right) contributions to the heat deposition in the actual HTS materials within the core (excluding heat into the cold inconel center rod and the 316 L stainless steel inner pressure vessel included in **Figure 16**). The figure shows that the direct neutron energy deposition is still dominated by the "direct" contributions, principally from neutrons generated within the plasma itself, but also from neutrons scattered from the tungsten and water reflector layers beyond the plasma. The next highest contribution is around 20 times lower and come from n (n', gamma) interactions in the HTS material itself. In contrast, the principal gamma heating arises from (n, gamma), (n, n' gamma), (n,2n), and fluorescence reactions.

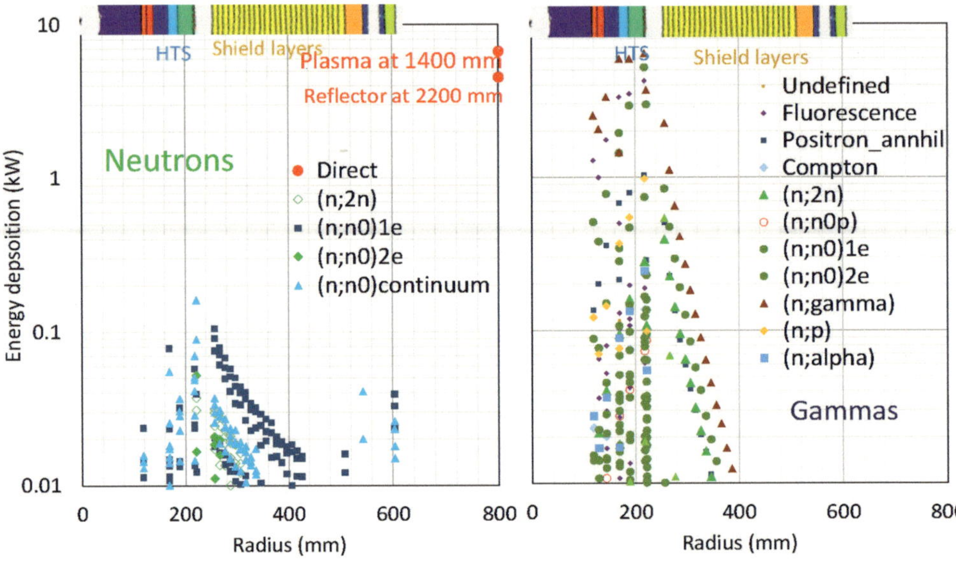

Figure 16.
The originating positions and reactions responsible for the neutron energy deposition into the HTS core volumes of a spherical tokamak power plant. The colored upper band and labels record the various tally volumes contributing to the energy deposition. The different symbols record the reaction method. "Direct" refers to neutrons originating in the plasma core itself, most from the plasma, some from the outer reflector, both off the scale of the neutron graph.

These tend to be largest close to and fall off quite rapidly into the shield layers. Direct gamma heating is negligible.

The effects of these results on radiation damage are clear. Fast neutron damage will arise from the 14-MeV neutron fluence component which penetrates the shield. Gamma damage in contrast can be identified as coming from (n, n' gamma) reactions in ^{63}Cu from the heat-sink material surrounding the HTS tapes and from ^{56}Fe and ^{58}Ni in the Hastalloy of the HTS tapes. (n, gamma) reactions from ^{184}W and ^{182}W in the shield are significant but reduce rapidly into the shield.

6. Conclusions

Spherical tokamaks with HTS magnets are one of few options for safe, abundant energy on a small footprint. This chapter addresses a key issue: finding a material that can shield the HTS core while maintaining small major radius and low aspect ratio. It explores the operational performance of tungsten boride tokamak shields particularly using the tally tagging option of MCNP to detail the locations, atoms, and reactions contributing to neutron and gamma fluence and energy depositions. These studies have revealed that the W_2B_5 shield operation is dominated by the slightly moderated near 14 MeV neutrons produced in the plasma. At all positions well within the shield, this fluence of neutrons is attenuated by inelastic (n, n' gamma) reactions to produce lower-energy neutrons which can be absorbed by ^{10}B and gammas which are absorbed by tungsten in a few centimeters. Our HPC capability has been instrumental in our design process. It will be increasingly used in our quest for faster fusion.

Acknowledgements

The authors are most grateful to their colleagues at Tokamak Energy, particularly Jack Astbury, Guy Morgan, Chris Wilson, Sandeep Irukuvarghula, Greg Brittles, Tony Langtry, Peter Buxton, Robert Slade and George Smith for their contributions to this work.

Author details

Colin Windsor* and Gurdeep Kamal
Tokamak Energy Ltd., Milton Park, Oxon, UK

*Address all correspondence to: colin.windsor@tokamakenergy.co.uk

References

[1] Windsor C. Can the development of fusion energy be accelerated? 2019. DOI: 10.1098/rsta.2017.0446

[2] Lawson J. Harwell report A.E.R.E GP/R 1807. 1955. Available from: https://www.euro-fusion.org/fileadmin/user_upload/Archive/wp-content/uploads/2012/10/dec05-aere-gpr 1807.pdf

[3] Costley AE, Hugill J, Buxton PF. On the power and size of tokamak fusion pilot plants and reactors. Nuclear Fusion. 2015;**55**:033001

[4] Costley AE. Towards a compact spherical tokamak fusion pilot plant. Philosophical Transactions of the Royal Society A. 2019;**377**:20170439. DOI: 10.1098/rsta.2017.0439

[5] Flyvbjerg B. What you should know about megaprojects and why: An overview. Project Management Journal. 2014;**45**(2):6-19,

[6] Sykes A, Turner MF, Patel S. Proceedings of the 11th European Conf. on Controlled Fusion and Plasma Physics. Vol. 2. Aachen, Germany; 1983. p. 363

[7] Peng Y-KM, Strickler DJ. Features of spherical torus plasmas. Nuclear Fusion. 1986;**26**:769

[8] Gryaznevich M et al. Achievement of record β in the START spherical tokamak. Physics Review Letters. 1998;**80**:3973-3975

[9] Bickerton RJ, Connor JW, Taylor JB. Diffusion driven plasma currents and bootstrap tokamak. Nature Physical Science. 1971;**229**:110

[10] Zarnstorff MC et al. Bootstrap current in TFTR. Physical Review Letters. 1968;**60**:1306

[11] Stork D. The upgrade to the Mega Amp Spherical Tokamak, International Atomic Energy Agency, Vienna (Austria); 637 p; Oct 2010; p. 473; FEC 2010: 23. IAEA Fusion Energy Conference; Daejeon (Korea, Republic of); 11-16 Oct 2010; ICC--P5-06

[12] National Spherical Torus Upgrade, Princeton plasma physics laboratory, USA. 2022. Available from: https://www.pppl.gov/research/nstx-u

[13] Globus M2 tokamak, Ioffe Institute, St. Petersberg, Russia. 2020. Available from: https://www.nature.com/articles/d42473-021-00131-5

[14] Tokamak Energy, Milton Park, Didcot, UK. 2022. Available from: https://www.tokamakenergy.co.uk/technology/st40/

[15] Bednorz JG, Müller KAZ. Possible high Tc superconductivity in the Ba-La-Cu-O system. Zeitschrift für Physik B Condensed Matter. 1986;**64**:189-193. DOI: 10.1007/BF01303701

[16] Molodyk A et al. Development and large volume production of extremely high current density YBa2Cu3O7 superconducting wires for fusion. Scientific Reports. 2021;**11**:11. DOI: 10.1038/s41598-021-81559-z

[17] Nuclear Engineering International News 23/09/2021 Tokamak Energy demonstrates new magnet technology for fusion power

[18] Tokamak Energy. Tests patented partial insulation technology. Superconductor Week. 2021;**35**(9)

[19] Hender JC, Wesley C, Bialek J, et al. Chapter 3: MHD stability, operational limits and disruptions.

Nuclear Fusion. 2007;**47**(6):S128. DOI: 10.1088/0029-5515/47/6/S03

[20] Windsor CG, Morgan JG. Neutron and gamma flux distributions and their implications for radiation damage in the shielded superconducting core of a fusion power plant. Nuclear Fusion. 2017;**57**:116032. DOI: 10.1088/1741-4326/aa7e3e

[21] Windsor CG, Astbury JO, Davidson JJ, McFadzean CJR, Morgan JG, Wilson C, et al. Tungsten boride shields in a spherical tokamak fusion power plant. Nuclear Fusion. 2021;**61**:086018. DOI: 10.1088/1741-4326/ac09ce

[22] Windsor CG, Astbury JO, Morgan JG, Wilson C, Humphry-Baker SA. Activation and transmutation of tungsten boride shields in a spherical tokamak. Nuclear Fusion. 2022;**62**:036009. DOI: 10.1088/1741-4326/ac4866

[23] Brookhaven National Nuclear Data Center, Evaluated nuclear data file

[24] McKinney GW. Tally tagging feature in MCNPX 2.7.A, Los Alamos report 999LA-UR-08-06930. 2008. Available from: https: //mcnp. lanl.gov/pdf_files/la-ur-08-6930.pdf

Chapter 6

Some Research Method about Superconducting Magnet Systems of TOKAMAK

Jian Rong

Abstract

The superconducting magnet operates in conditions of complex electromagnetics, which could cause hysteresis loss and coupling losses, the so-called AC losses. In this chapter, the AC losses calculation of superconductor will be discussed in detail. Usually, the superconducting magnets are wound by superconducting coils, which are twisted by superconducting wires. The length of superconducting wires is hundreds of meters, while the length of coils is millions of meters; thus, joints are needed to join the coils. The design of different patterns of joints, such as twin-box joint, butt joint, and petal overlap joint, will be introduced in detail. Joule heat and AC losses in the joint may cause locality quench, and if the design stability margin of the magnet could not cover the joule heat and losses, the locality quench will cause global quench of the magnet. The temperature rise caused by joule heat and AC losses will be discussed in detail. Furthermore, the magnetic Lorentz force and mechanical displacement could cause locality quench, which may cause a global quench, once the coolant could not take away the heating pulse. The simulation of the stability and quench behavior of the superconducting cable-in-conduit conductor will be introduced in detail.

Keywords: AC losses, hysteresis loss, coupling losses, superconducting joint, current sharing temperature, temperature margin, minimum quench energy, hotspot temperature

1. Introduction

The TOKAMAK (toroidal, kamera, magnet, kotushka) is a typical magnetic confinement fusion device, which uses a powerful magnetic field to confine plasma in the shape of a torus. Long-time plasma confinement is needed for commercial electricity of tokamak, the superconducting magnet system could provide a long-time stable confinement field for the plasma of tokamak, and the plasma could operate a long time stably. Thus, the superconductivity is considered one of the key technologies to reach commercial electricity for tokamak.

There are two key components (superconducting CICC (cable-in-conduit conductor) and superconducting joints) in the superconducting magnet system. Superconducting CICC design refers to twist pattern, superconducting filament,

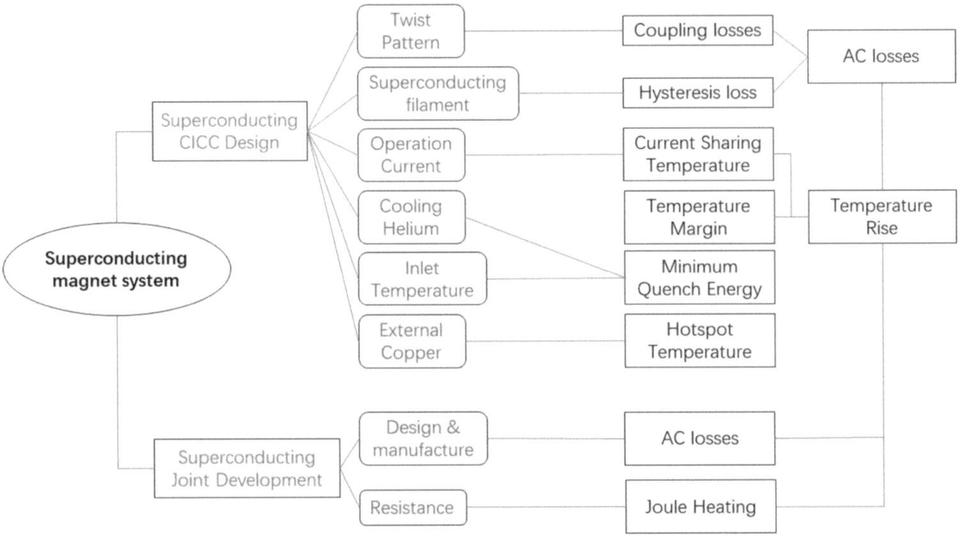

Figure 1.
The chapter structure diagram.

operation current, cooling helium, inlet temperature, external copper, and so on, while superconducting joint design refers to the design, manufacture, and resistance, as shown in **Figure 1**. In this chapter, this factor of the CICC and joint design will be introduced in detail.

In the complex varying electromagnetic environment, the changing magnetic field could produce AC losses (include hysteresis loss and coupling losses) in the superconducting conductor and joints, which could cause a quench. The coupling losses are affected by the twisted pattern of the conductor, while the superconducting filament decides the hysteresis loss. AC losses and joint resistance will cause a temperature rise, which reduces the temperature margin, the cooling helium and inlet temperature decide the minimum quench energy, and the hotspot temperature is decided by the external copper. In this chapter, AC losses, current sharing temperature, and joint resistance calculation method will be introduced in detail, taking ITER TF (Toroidal Field) main busbar as an example, and its quench and stability behavior in 15 MA plasma current scenario will be discussed, which includes the temperature rise caused by AC loss and joint resistance, minimum quench energy, hotspot temperature. And a joint development design is introduced in detail, hoping that this chapter will help reader know how to study superconducting magnet system.

2. AC losses calculation of the ITER TF main busbar

A feeder system is used to supply the electrical power, cryogens, and control system interfaces outside the cryostat through a warm-cold barrier to the ITER magnets. Each feeder consists of control cubicles, Dry Box (DB), Coil Terminal Box (CTB), Pressure Release Valve Rack (PRVR), S-Bend Box (SBB), Cryostat Feed Through (CFT), and In-Cryostat Feeder (ICF) [1], as shown in **Figure 2**. The main busbar of the TF feeder is a kind of typical superconducting CICC, which employs 900 Ni-plated NbTi strands and 522 Ni-plated copper strands, and the pattern of the main

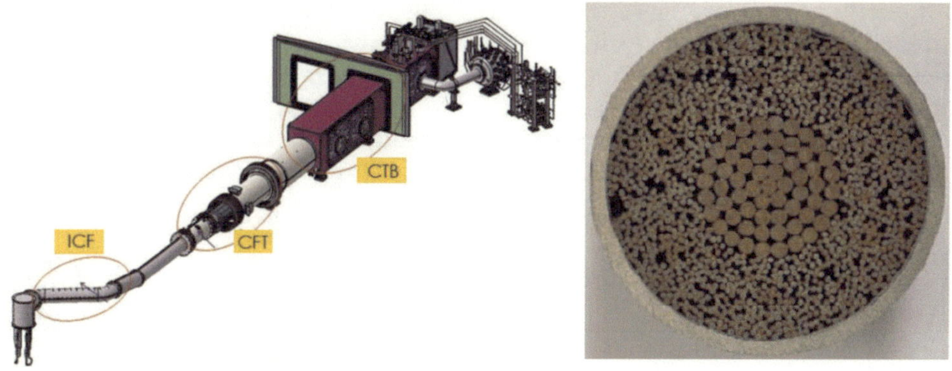

Figure 2.
The feeder system of the TF magnet of the ITER [2].

busbar conductor is (2SC + 1Cu) × 3 × 5 × (5 + C0) × (6 + C1), where C0 is 3 × 4Cu, and C1 is (1 + 6 + 12 + 18 + 24)Cu, and the sub-cables show petal structure without wraps. The complex changing complex electric-magnetic causes AC losses in the main busbar conductor. In this section, the AC losses calculation of the main busbar will be introduced in detail.

2.1 The hysteresis loss calculation of ITER main busbar

In the varying magnetic field, the moving flux will deposit the magnetic energy in the superconductor so-called hysteresis loss. Because there is flux pinning in the superconductor, the flux overcomes the pinning potential energy and surface potential barrier and does work, thus the hysteresis loss is produced. There are following four main kinds of calculation methods of hysteresis loss [3]:

a. In the varying magnetic field, there is an induced current in the superconductor, the hysteresis loss could be calculated by calculation of energy flowing into and out of the superconductor, and the hysteresis loss equals the integration of the Poynting vector ($S = E \times H$) on the field loop, where the electrical field could be calculated by Maxwell's equation of $\nabla \times E = -dB/dt$.

b. The hysteresis loss could be calculated by integrating the point multiplication of current density and electrical field ($J \cdot E$).

c. The hysteresis loss could be calculated by integrating the magnetizing curve ($Q = M(H)dH$) in the varying magnetic field.

d. The hysteresis loss per unit volume equals the work of Lorentz force ($SP = (J \times E) \cdot v$), where the v is the velocity of movement of the flux.

In the superconducting strands, the pattern of superconductor is twisted superconducting filaments, then the second method is chosen to calculate the hysteresis loss of superconductor. The superconductor is equivalent into a cylinder, and the Bean's critical state model is used in hysteresis loss calculation [4, 5], in which the field flux penetrates the superconductor fully, and the current density is critical current density of superconductor. According to classical electromagnet theory, hysteresis loss per unit volume equals the product of electrical field and critical current density ($Jc{\cdot}E$). In the condition that the background field flux penetrates the superconductor fully, the field direction is perpendicular to the axis of the cylindrical superconductor, and the electrical field direction is parallel to Z-axis, as shown in **Figure 3**, H_θ and H_r stand for the component in the direction of $\vec{\theta}$ and $\vec{r}$ of magnet field intensity, respectively.

According to Maxwell's equation, the following equation could be given [6]:

$$\nabla \times \vec{E} = -\mu_0 \frac{\partial \vec{H}}{\partial t} = \begin{vmatrix} \vec{e_r} & \vec{e_\theta} & \vec{e_z} \\ \frac{\partial}{\partial r} & \frac{1}{r}\frac{\partial}{\partial \theta} & \frac{\partial}{\partial z} \\ 0 & 0 & E_z \end{vmatrix} \rightarrow \begin{matrix} \frac{1}{r}\frac{\partial E_z}{\partial \theta} = -\mu_0 \frac{\partial H_r}{\partial t} \\ \frac{\partial E_z}{\partial r} = -\mu_0 \frac{\partial H_\theta}{\partial t} \end{matrix} \quad (1)$$

Where E_z equals $\left|\vec{E}\right|$, is the component of $\vec{E}$ in the Z-axis direction. Once H_r and H_θ are replaced by H, the Eq. (1) could be turned into the following equation:

$$\begin{matrix} \frac{1}{r}\frac{\partial E_z}{\partial \theta} = -\mu_0 \frac{\partial H}{\partial t} sin\theta \\ \frac{\partial E_z}{\partial r} = -\mu_0 \frac{\partial H}{\partial t} cos\theta \end{matrix} . \quad (2)$$

Solve the Eq. (2), the following equation could be given:

$$E_z = \mu_0 \frac{dH}{dt} rcos\theta = \frac{dB}{dt} rcos\theta, \quad (3)$$

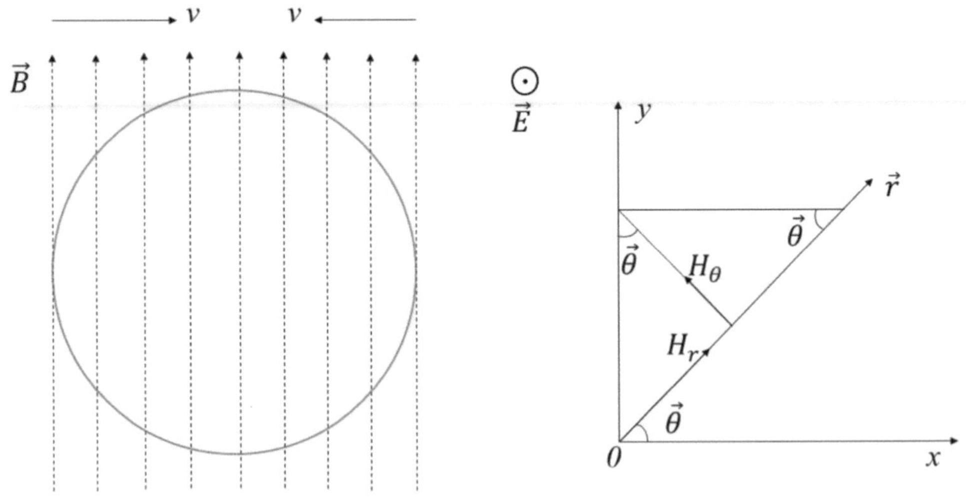

Figure 3.
The flux penetrates the cylindrical superconductor in the varying magnetic field, where v is the velocity of the flux moving into and out of the superconductor.

Then the hysteresis loss per unit length could be got in a polar coordinates system:

$$P = \int_S \vec{J} \times \vec{E} dS = 4\int_0^{\frac{d_f}{2}} J_c \frac{dB}{dt} r^2 cos\theta drd\theta = \frac{1}{6} J_c \frac{dB}{dt} d_f^3. \tag{4}$$

Then the hysteresis loss per unit volume is given as the following (W/m^3):

$$P = \frac{2}{3\pi} \frac{dB}{dt} J_c d_f \tag{5}$$

Then the hysteresis loss of superconducting CICC per unit length is carried out by the following equation (only the loss when the magnetic field is perpendicular to the axis of a superconductor is considered) (W/m):

$$P_h = \frac{2}{3\pi} \frac{dB}{dt} J_c d_f A_{sc}, \tag{6}$$

Where *Jc* is the critical current density of the superconductor, d_f is the effective diameter of the superconducting filament, A_{sc} is the total area of the superconductor, and *dB/dt* is the variation of the magnetic field. The effective diameter of a superconducting filament is affected by the connection between filaments, and the effective diameter is determined by experiment only, the effective diameter of a superconducting filament equals the average diameter of the filament during hysteresis loss calculation.

There are superconductors and normal conductors in the CICC, and because of hysteresis loss, it is represented that there is small resistance in the CICC, thus there is a small current in the normal conductor. Therefore, if the critical current density of the CICC is higher than *Jc*, then the hysteresis loss is given as the following Eq. (W/m):

$$P_h = \frac{2}{3\pi} J_c \left(1 + \left(\frac{I}{J_c A_{sc}}\right)^2\right) d_f A_{sc} \frac{dB}{dt}, \tag{7}$$

Where *I* is the current of the superconducting CICC.

The critical current density of NbTi superconductor is calibrated as the function of temperature and background field by Bottura fie [7, 8], shown as the following function:

$$\begin{gathered} J_c(B,T) = \frac{C_0}{B}(b)^{0.8}(1-b)^{1.2}\left(1-t^{1.7}\right)^{2.61} J_{cre} \\ b = \frac{B}{B_{c20}(1-t^{1.7})}, \\ t = \frac{T}{T_{c0}} \end{gathered} \tag{8}$$

Where C_0 is the normalization constant of 43.125 T of NbTi, B_{c20} is the upper critical magnetic field flux density with the temperature of 0 K, T_{c0} is the critical temperature with the magnetic field of B = 0, and J_{cre} is the normalized value of critical current density at field 5.0 T and temperature 4.2 K (2800 A/mm^2) in the condition of NbTi superconductor.

The parameters of TF main busbar CICC of ITER are listed in **Table 1**, the diameter of the superconducting filaments is 8 μm, and the total cross-sectional area of the superconductor is 111.2 mm^2. In the reference scenario of 15 MA plasma current, the maximum magnetic field of the main busbar is 1.887 T and the minimum is 0.045 T, and the maximum variation of the magnetic field is 0.18 T/s. According to Eq. (7), the hysteresis loss is carried out, and the maximum value is 0.185 W/m. There is induced

Parameters		**Value**	**Unit**
Current	I	68	kA
Ratio of Cu to no-Cu of the strand	λ	2.35	mm^2
Radius of copper core zone of strand	r_c	0.175	mm
Radius of filamentary zone of strand	r_f	0.313	mm
Radius of copper shell zone of strand	r_{ms}	0.363	mm
Radius of the strand	R_0	0.365	mm
The diameter of the superconducting filament	d_f	8	μm
Thickness of Nickel plating	e_b	2	μm
Residual Resistivity Ratio of Ni and copper	RRR	100	
Cable twist pitch: (strand)	p_0	15	mm
1st	p_1	45	mm
2nd	p_2	85	mm
3rd	p_3	145	mm
4th	p_4	250	mm
5th	p_5	450	mm
Number of superconducting strands	N_{SC}	900	
Number of copper strands	N_{Cu}	522	
Diameter of Copper-core C1	r_5	17.73	mm
Inner conduit diameter	R_5	41.0	mm
Stainless steel conduit thickness		2.0	mm
Superconducting cross section	A_{SC}	111.2	mm^2
Copper cross section	A_{Cu}	671.0	mm^2
The bundle helium cross section	A_{He}	416.32	mm^2
The central helium cross section	A_{cen}	49.38	mm^2
C1 void fraction	ϑ_{C1}	20.0	%
Bundle void fraction	ϑ_{bundle}	38.78	%
Total void fraction of the conductor	ϑ_{busbar}	35.2	%

Table 1.
Properties of the TF main busbar of ITER [9–11].

current in the copper strands and copper matrix, and the flowing induced current causes coupling losses in the main busbar, which will be discussed in detail in the next section.

2.2 The coupling losses calculation of the ITER TF main busbar

In the superconducting CICC, there are normal conductor strands, a normal conductor matrix, and high-resistance plating, in which the induced current flowing causes the coupling losses. During coupling losses calculation, the superconducting strand is treated as a homogeneous mixture, and the magnetic field is assumed uniform distribution in the superconducting strands [7].

In the varying magnetic field, there is a coupling current in the matrix, because of the twist of superconducting filament, and the twist pitch is much longer than the diameter of superconducting filaments, the direction of the coupling current is parallel to the background field, and is different to the normal eddy current, as shown in **Figure 4**.

As shown in **Figure 4**, the total area surrounded by twisted superconducting filament in the Y-Z plane is given by following equation:

$$S(z) = \int_0^z R_f \cos\left(\frac{2\pi x}{p}\right)dx = \frac{p}{2\pi} R_f \sin\left(\frac{2\pi z}{p}\right). \tag{9}$$

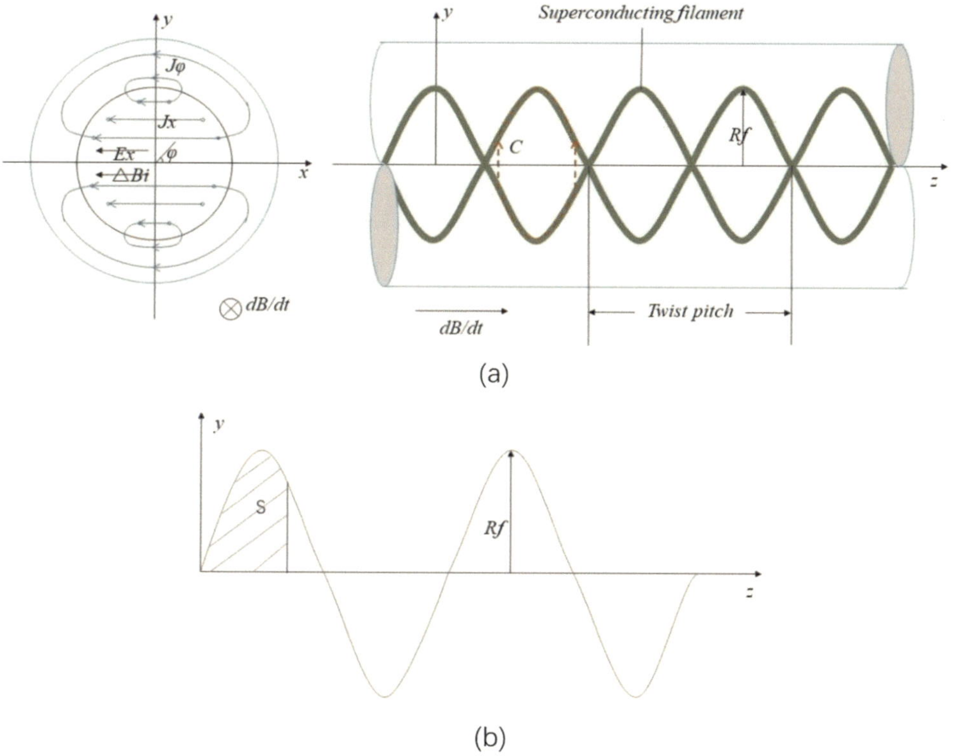

Figure 4.
The induced current in the superconducting filament and the area surrounded by the superconducting filament. (a) the induced current of the superconducting filament. (b) the area surrounded by the superconducting filament.

The induced electromotive force surrounded by C circuit in the Z direction, as shown in **Figure 4**, is given by following equation:

$$V(z) = \frac{-d\varnothing}{dt} = -S\frac{dB_i}{dt} = \frac{-p}{2\pi}R_f \sin\left(\frac{2\pi z}{p}\right)\frac{dB_i}{dt}, \tag{10}$$

Where p is the twist pitch of the superconducting filament, R_f is the radius of the twist zone, and the *dBi/dt* is the variation of the background magnetic field. Similarly, the induced electromotive force of X-direction in the cross section of superconducting filament could be determined by the following equation:

$$V(x) = \frac{-p}{2\pi}x\frac{dB_i}{dt}. \tag{11}$$

Then the average induced electric field strength is given by following equation:

$$\overrightarrow{E_x} = \frac{-p}{2\pi}\frac{dB_i}{dt}. \tag{12}$$

Then the transverse induced current density is given by following equation:

$$J_x = \frac{-p}{2\pi\rho_t}\frac{dB_i}{dt}. \tag{13}$$

Thus the transverse induced electric field strength is calculated by following equation:

$$E_x = J_x\rho_t = \frac{-p}{2\pi}\frac{dB_i}{dt}, \tag{14}$$

Where ρ_t is the transverse resistivity of the superconducting filament zone, and it is assumed that e is the thickness of the high resistivity layer around the superconducting filaments. The induced electrical field induces a current, whose direction is parallel to the induced electrical field, the current has an azimuth angle, and flows through a matrix. If $e \ll R_f$, the current density in an angular direction, as shown in **Figure 4**, is:

$$J_\varphi = \frac{-p}{2\pi\rho_n}\frac{dB_i}{dt}\sin(\varphi), \tag{15}$$

Where the ρ_n is the resistivity of the matrix out of the superconducting filaments zone, and φ is the angle between the axis of superconducting filaments and the background field shown in **Figure 4**.

The current density in the surface of superconducting filaments could be given by the correct balance current of transverse and angular induced currents:

$$J_x = \left(\frac{p}{2\pi}\right)\left(\frac{1}{\rho_t} + \frac{e}{R_f\rho_n}\right)\frac{dB_i}{dt}\sin(\varphi). \tag{16}$$

The coupling losses caused by the induced current flowing in the matrix of the superconducting strand is $P_c = J_x E_x$, generally, the angular φ equals to $\pi/2$, therefore, the coupling losses of superconducting filaments zone are calculated by the following equation:

$$P_c = \frac{p}{2\pi}\left(\frac{1}{\rho_t} + \frac{e}{R_f\rho_n}\right)\frac{dB_i}{dt}\left(\frac{-p}{2\pi}\frac{dB_i}{dt}\right) = \left(\frac{1}{\rho_t} + \frac{e}{R_f\rho_n}\right)\left(\frac{p}{2\pi}\right)^2\left(\frac{dB_i}{dt}\right)^2. \quad (17)$$

A physical parameter that the coupling time constant of superconducting filament zone is defined:

$$\tau = \frac{\mu_0}{2}\left(\frac{1}{\rho_t} + \frac{e}{R_f\rho_n}\right)\left(\frac{p}{2\pi}\right)^2. \quad (18)$$

Because $e \ll R_f$, then relative $1/\rho_t$, $e/R_f\rho_n$ could be ignored, then the coupling time constant and coupling losses are shown by following equation:

$$\tau = \frac{\mu_0}{2\rho_t}\left(\frac{p}{2\pi}\right)^2, \quad (19)$$

$$P_c = \frac{2\tau}{\mu_0}\left(\frac{dB_i}{dt}\right)^2. \quad (20)$$

And the coupling losses per unit length superconducting conductor are shown by following equation:

$$P_c = \frac{2\tau}{\mu_0}\left(\frac{dB}{dt}\right)^2(1+\lambda)A_{sc}, \quad (21)$$

Where λ is the ratio of matrix and superconductor.

Generally, the superconducting strands are the composite structure of the normal conductor matrix and superconductor, the superconducting filaments twists in the strands immerse in the matrix. The superconducting strands consist of a normal conductor matrix, superconducting filament, a high resistivity layer, and a sheath around the strands, in which the superconducting filaments distribute evenly. The cross sections of superconducting strands are shown in **Figure 5**, and the cross-section

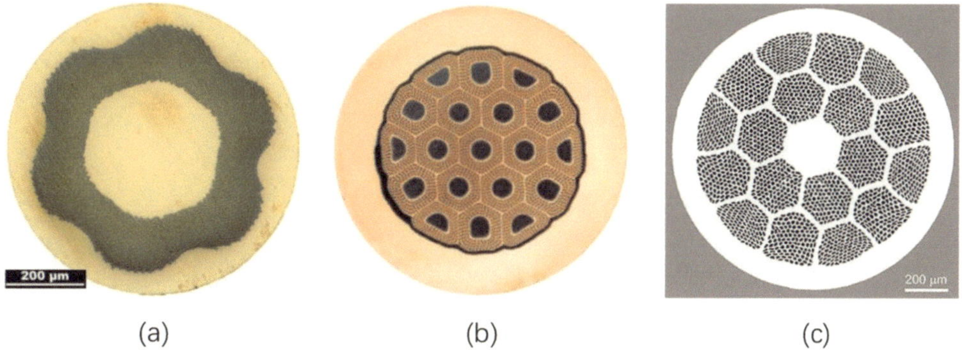

Figure 5.
(a) The cross section of NbTi strand of ITER main busbar [9], (b) the cross section of Nb3Sn strand of ITER TF coils (OST) [10], (c) the cross section of Bi-2212 round wire [12].

strand is divided into the core zone, superconducting filamentary zone, the matrix shell zone, high-resistivity layer, and sheath of the strands.

The cross section of superconducting strands is equivalent to several concentric circles during coupling losses of the strand calculation, the total coupling time constant of the strand is the sum of the time constant of all zones, as shown in the following equation:

$$\tau_0 = \tau_c + \tau_f + \tau_{mf} + \tau_b + \tau_{ms} + \tau_e, \tag{22}$$

Where τ_0, τ_c, τ_f, τ_{mf}, τ_b, τ_{ms}, and τ_e are the total time constant of the total strand, the time constant of the core zone, filamentary zone, filamentary shell zone, high resistivity layer, sheath of the strands, and the eddy current, respectively. According to Eq. (19), the time constant of each zone is given as follows [7, 13]:

a. The time constant of core zone:

$$\tau_c = \frac{\mu_0}{2\rho_{mc}}\left(\frac{p}{2\pi}\right)^2 \frac{r_c^2}{r_{ms}^2}, \tag{23}$$

b. The time constant of filamentary zone:

$$\tau_f = \frac{\mu_0}{2\rho_{tf}}\left(\frac{p}{2\pi}\right)^2 \frac{r_f^2 - r_c^2}{r_{ms}^2}, \tag{24}$$

c. The time constant of filamentary shell zone:

$$\tau_{mf} = \frac{\mu_0}{2\rho_{mf}}\left(\frac{p}{2\pi}\right)^2 \frac{r_{mf}^2 - r_f^2}{r_{ms}^2}\left(r_f^2 + b_1^2 r_{mf}^2\right)a_1^2, \tag{25}$$

d. The time constant of high-resistivity layer:

$$\tau_b = \frac{\mu_0}{2\rho_b}\left(\frac{p}{2\pi}\right)^2 \frac{r_f^2}{r_{mf}^2}\frac{r_b^2 - r_{mf}^2}{r_{ms}^2}\left(r_{mf}^2 + b_2^2 r_b^2\right)a_2^2, \tag{26}$$

e. The time constant of the sheath:

$$\tau_{ms} = \frac{\mu_0}{2\rho_s}\left(\frac{p}{2\pi}\right)^2 \frac{r_{mf}^2}{r_b^2}\frac{r_{ms}^2 - r_b^2}{r_{ms}^2}\left(r_{ms}^2 + r_b^2\right)a_3^2, \tag{27}$$

f. The time constant of the eddy current:

$$\tau_e = \frac{\mu_0}{8}\left(\frac{r_c^2}{\rho_{mc}} + \frac{\delta\left(r_f^4 - r_c^4\right)}{r_f^2\rho_{tf}} + \frac{r_{mf}^4 - r_f^4}{r_{mf}^2\rho_{mf}} + \frac{r_b^4 - r_{mf}^4}{r_b^2\rho_b} + \frac{r_{ms}^4 - r_b^4}{r_{ms}^2\rho_{ms}}\right), \tag{28}$$

Where the r_c, r_f, r_{mf}, r_b, and r_{ms} are the radius of core zone, filamentary zone, filamentary shell zone, high resistivity layer, and sheath of the strands, respectively; ρ_c,

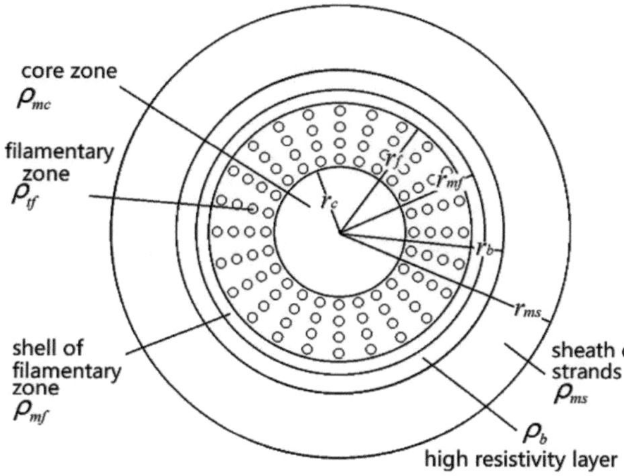

Figure 6.
The schematic layout of the cross section of the superconducting strand.

ρ_{mf}, ρ_b, and ρ_{ms} are the resistivity of core zone, filamentary shell zone, high-resistivity layer, and sheath of the strands respectively, as shown in **Figure 6**, and the δ is the ratio of the superconductor in the filamentary zone. In the previous equation, a_1, a_2, a_3, b_1, and b_2 are given by the following:

$$
\begin{aligned}
a_1 &= \frac{r_f}{r_f^2 + b_1 r_{mf}^2} \\
a_2 &= \frac{a_1(1+b_1)r_{mf}^2}{r_{mf}^2 + b_2 r_b^2} \\
a_3 &= \frac{a_2(1+b_2)r_b^2}{r_b^2 + r_{ms}^2} \\
b_1 &= \frac{\rho_b\left(r_{mf}^2 + b_2 r_b^2\right) + \rho_{mf}\left(b_2 r_b^2 - r_{mf}^2\right)}{\rho_b\left(r_{mf}^2 + b_2 r_b^2\right) - \rho_{mf}\left(b_2 r_b^2 - r_{mf}^2\right)} \\
b_2 &= \frac{\rho_{ms}\left(r_{ms}^2 + r_b^2\right) + \rho_b\left(r_{ms}^2 - r_b^2\right)}{\rho_{ms}\left(r_{ms}^2 + r_b^2\right) - \rho_b\left(r_{ms}^2 - r_b^2\right)}
\end{aligned}
\quad . \tag{29}
$$

ρ_f and ρ_{tf} are the effective resistivity and equivalent resistivity, respectively, which are given by the following equation:

$$
\begin{aligned}
\rho_{tf} &= \frac{(1-\lambda)\rho_m + (1+\lambda)\rho_f}{(1+\lambda)\rho_m + (1-\lambda)\rho_f}\rho_m \\
\rho_f &= \frac{2R_{cmf}}{d_f}
\end{aligned}
\quad , \tag{30}
$$

Where ρ_m is the resistivity of the matrix in the filamentary zone, R_{cmf} is the contact resistance of filament-to-matrix, and λ is the ratio of a superconductor to the matrix in the superconducting filament zone, which is given by the following equation:

$$\lambda = \frac{A_f}{\pi\left(r_f^2 - r_c^2\right) - A_f}, \tag{31}$$

Where A_f is the cross-sectional area of the filamentary zone.

According to the need for cooling, support, and steady state, the magnet system of tokamak uses a CICC, in which the conductor is multistage cabled by superconducting and copper strands in a conduit carrying single-phase supercritical helium.

The total time constant of the conductor equals the sum of all contributions time constant [7]:

$$\tau = \sum_{n=0}^{N} \tau_n, \tag{32}$$

Where τ_0 is the time constant of the superconducting strand, τ_n is associated with the contribution of each cabling stage sub-cable ($n \geq 1$), which is calculated by following method:

$$\tau_n = \frac{\mu_0}{2\rho_n}\left(\frac{p_n}{2\pi}\right)^2 \frac{1}{1 - \vartheta_n}, \tag{33}$$

where p_n, ρ_n, and ϑ_n are respectively the effective twist pitch length, the effective resistivity, and the average void fraction of cabling stage n. The effective resistivity ρ_n refers to the contact resistance and plating resistance of the contact section, then p_n and ρ_n are calculated by following equations:

$$\begin{aligned} p_n &= p_n - \frac{r_{n-1}}{R_{n-1}} p_{n-1} \\ \rho_n &= \frac{\rho_b e_b}{\varepsilon_n R_{n-1}} + R_{contact} R_{n-1} \end{aligned}, \tag{34}$$

where p_n is the cabling twist pitch for full back twist, ε_n is the contact area ratio, R_n and r_n are the outer and the twist radius in stage n, respectively, which were calculated by geometrical method [14]. and r_0 and R_0 were the filamentary zone radius and round wire radius, respectively. The contact resistance $R_{contact}$ of inter-strands and inter-sub-cables is defined with unit of Ω, where ρ_b and e_b are the material resistivity and the thickness of the plating, respectively. ε_n is the contact area ratio of each stage sub-cables, the contact area ratio and the radius of sub-cable have following relations:

$$\varepsilon_n = \frac{R_{n-1}}{R_n} \varepsilon_{n-1}. \tag{35}$$

This formula is accurate in the range of the void fraction between 28 and 40%, and it can be known that the ratio of contact area increases as the radius of the twisted cable decreases. In this book, Dr. Rong introduces a method to calculate the contact area ratio by the geometrical method.

The micrograph cross section and simplified diagram of the NbTi superconducting strands are shown in **Figure 7** [15], the strand cross section is simplified into three parts such as the copper core zone, superconducting filamentary zone, and copper shell zone, whose radii are measured by the simplified diagram, listed in **Table 1**. In

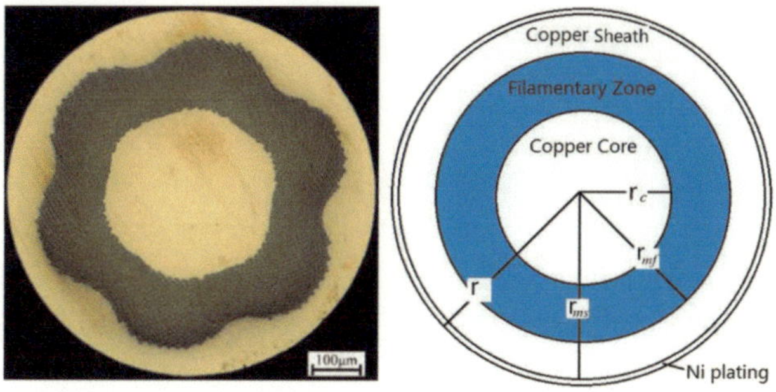

Figure 7.
The cross section and schematic layout of the superconducting strand of the ITER main busbar.

the superconducting filamentary zone of the busbar strands, there is no high resistance layer around the superconducting filaments and a high resistivity layer zone, then the radius of the filamentary shell zone equals the radius of the high resistivity layer, which all equals the radius of the filamentary zone, meaning that $r_{mf} = r_b = r_f$. Thus, the coupling losses of filamentary shell zone and high resistivity zone equals to 0, meaning that $\tau_{mf} = \tau_b = 0$, so the time constant τ_0 of the superconducting strands consists of following four contributions,

$$\tau_0 = \tau_c + \tau_f + \tau_{ms} + \tau_e, \tag{36}$$

where τ_c, τ_f, and τ_{ms} are the time constant of the copper core, filamentary zone, and the copper shell zone, respectively, and τ_e dues to eddy current. The matrix of the NbTi strands is copper.

$$\begin{aligned} \tau_c &= \frac{\mu_0}{2\rho_{Cu}}\left(\frac{p}{2\pi}\right)^2 \frac{r_c^2}{r_{ms}^2} \\ \tau_f &= \frac{\mu_0}{2\rho_{tf}}\left(\frac{p}{2\pi}\right)^2 \frac{r_f^2 - r_c^2}{r_{ms}^2} \\ \tau_{ms} &= \frac{\mu_0}{2\rho_{Cu}}\left(\frac{p}{2\pi}\right)^2 \frac{r_f^2}{r_{ms}^2} \frac{r_{ms}^2 - r_f^2}{r_{ms}^2 + r_f^2} \\ \tau_e &= \frac{\mu_0}{8\rho_{Cu}}\left(r_c^2 + \frac{\delta\left(r_f^4 - r_c^4\right)}{r_f^2} + \frac{r_{ms}^4 - r_f^4}{r_{ms}^2}\right) \end{aligned}, \tag{37}$$

Where ρ_{Cu} is the resistivity of the copper matrix of the superconducting strand, ρ_{tf} is the transverse resistivity in the filamentary zone, in which the contact resistance of filament-to-matrix R_{cmf} is 5.5 × 10–15 Ωm^2 [13]; and δ is the copper ratio of the filamentary zone, which could be calculated by the following:

$$\delta = \frac{1}{1 + \lambda}, \tag{38}$$

where λ is the ratio of the superconductor to the matrix in the superconducting filament zone.

Zone	Core	Filamentary	Shell	Due to eddy	Sum strand
$\tau(ms)$	6.94	38.72	3.62	1.65×10^{-4}	49.08

Table 2.
The coupling time constant of the superconducting strand.

The coupling time constant of the superconducting strand of the TF main busbar of ITER is calculated, and the total coupling time constant of the strand is 49.08 ms, as shown in **Table 2**.

The TF main busbar of ITER is wound from multiple stage sub-cables made with cooper and superconducting strands, and there is inducing current flow between adjacent strands and adjacent sub-cables, so there are coupling losses in sub-cables. All strands, both superconducting and copper strands, are Ni-plated to induce AC losses and to guarantee stable contact resistance with time. The cable pattern (strand as the first stage) of the TF-MB is (2SC + 1Cu) × 3 × 5 × (5 + C0) × (6 + C1), where C0 is 3 × 4Cu, and C1 is (1 + 6 + 12 + 18 + 24)Cu, as shown in **Figure 2**. The main busbar cable is simplified to five stages sub-cable for coupling time constant calculation, as shown in **Figure 8**, R_n and r_n are respectively the outer and the twist radius of the *nth* stage sub-cables, which carried out by a geometric method according to **Figure 8**, and r_0 is the strand filamentary area radius of the superconducting strand, while R_0 is the radius of the strand.

The effective resistivity of the sub-cables refers to the thickness and resistivity of Ni-plating and contact resistance, and the contact resistance is affected by the loading cycles. The contact resistance inter-strands and inter-sub-cables are tested in SULTAN [16–18] (the length of the simple is 0.5 m), where the Ni-plating petal has a peak contact resistance, and gradually decreases to 150 nΩ at cycle 30,000, and does not change much after 30,000 cycles. Despite the difference in applied load and conductor type, all measured conductors show inter-petal contact resistance around 150 nΩ after 30,000 loading cycles. And the inter-strands contact resistance is around 75 nΩ after 30,000 loading cycles. In the ITER feeder system, the main busbar is assembled by enough loading cycles burn-in conductor, so in 15MA Plasma Current Reference Scenario, the inter-strand contact resistance of 75 nΩ and inter-petal of 150 nΩ are chosen during coupling time constant calculation.

The contact area ratio is the ratio between the real contact and the surface area of the sub-cables, which is a key parameter for the coupling time constant. It is known

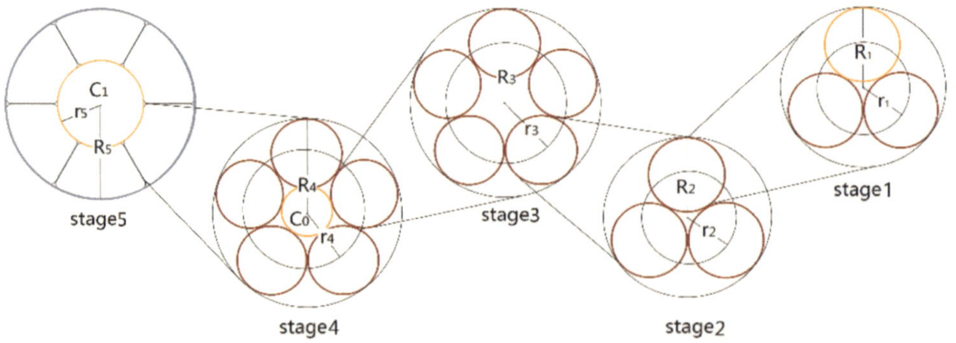

Figure 8.
The schematic view of the cabling geometry of the ITER TF main busbar.

that the contact area ratio of sub-cables could be calculated by Eq. (35), and the contact area ratio of the first sub-cable means that the contact area ratio between strands, is assumed a typical value of 20% [19]. However, contact area ratios of cables with different average void fractions are different, in the last stage sub-cable, the contact area ratio is quite low, because the pattern of the last stage sub-cable is petal structure, and there is wrap wrapped petals sometimes. In this chapter, a geometric method is introduced to estimate the contact area ratios between strands and of the last stage sub-cables, in which the contact angular thicknesses between two strands and between two sub-cables are the same. In the first stage, the contact area ratio could be estimated by the following equation:

$$\varepsilon_1 = \frac{l_{r1}}{l_{r2}}, \tag{39}$$

Where ε_1 characterizes the area of contact between two strands, l_{r1} refers to the contact line (red), and l_{r2} refers to the perimeter, meaning that the contact line (red) plus un-contact line (green), as shown in **Figure 9(a)**. In the ITER TF main busbar conductor, the first stage sub-cable consists of three green strands, and there is an indentation between contacted strands, as shown in **Figure 9(a)**.

In the bundle of the main busbar, if the cross section of the strands is standard circle, the bundle void fraction could be calculated by following equation:

$$\vartheta_{bundle} = 1 - \frac{(N_{SC} + N_{Cu})R_0^2\pi}{(R_5^2 - R_5^2)\pi}, \tag{40}$$

In the TF main busbar conductor, the calculated bundle void fraction with standard circle strand equals 44.55%, which is higher than the tested bundle void fraction of 38.78%, meaning that the strands are extruded to deformation, and it could be confirmed by the indentation in the untwisted strands. The indentation in the strand and the helium flow channel between adjacent strands are assumed same, then the

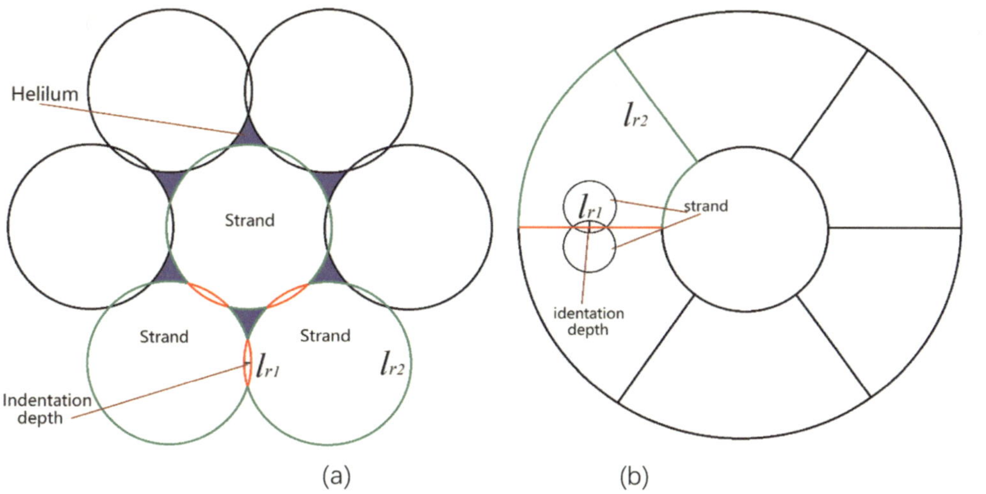

Figure 9.
The diagram of contact between Bi-2212 round wires (a), the diagram of contact between last stage sub-cables (petals) (b) for contact area ratio calculation.

indentation depth could be calculated by the bundle void fraction induction, as shown in **Figure 9(a)**, the blue area is the bundle helium flow channel.

In the last stage sub-cable of the TF main busbar conductor, the conductor is cabled by six petals, as shown in **Figure 9(b)**. The strands between two petals contact just like the red line of **Figure 9(b)**, then the contact area ratio between two petals, could be evaluated by Eq. (39). In which, l_{r1} refers to the contact line (red), and l_{r2} refers to the contact line (red) plus the un-contact line (green).

In 15MA plasma current reference scenario, the coupling time constant of the TF main busbar is carried out by Eq. (32), the coupling time constant after 30,000 loading cycles is calculated as 253.76 ms. The time constant of the cable is tested in SUTAN, $n\tau$ are about 160 ms after 10 cycles and 279 ms after 1000 cycles, respectively [20].

There are AC losses in the superconducting conductor, and also AC losses present in the superconducting joint, farther more, there is joule heating in the joint caused by the joint resistance. The AC losses and joule heating of the joints could cause a temperature rise in the main busbar, which reduces the temperature margin. In the next section, a development of petals overlap joint will be discussed in detail, the joint has lower resistance and AC losses, and small volume, which cause a lower temperature rise comparing with twin-box joint. And other improvements about the petals overlap joint are given, such as better coolant channels, more isotropous insulator stress around the joint.

3. The petal overlap superconducting joint for ITER TF main busbar

Typically, the superconducting busbar of the TOKAMAK feeder system is divided into several segments, which are connected by superconducting joints. In this section, a petal overlap superconducting joint for ITER TF main busbar will be discussed.

3.1 The design of the petal overlap superconducting joint

The current and coolant of the superconducting magnet are supplied by the Feeder, and for the purpose of assembly and mechanical redundancy, the feeder is divided into Coil-Terminal-Box (CTB), S-Bend-Box (SBB), Cryostat Feed-through (CFT), and In-Cryostat Feeder (ICF) [9, 21–23]. The following are the design requirements from ITER organization:

- The resistance of the joint should be lower than 2 nΩ at the condition of 4.5 K temperature, none background field, and carrying 68 kA current [24].

- After temperature rise caused by AC losses and resistance, the main busbar should have a temperature margin above 2.0 K [25, 26].

- In the joint, each strand of the cable should have ohmic contact with the copper sole, thus the copper sole of the joint should be longer than the twist pitches of the sub-cable contacted with it.

For lower resistance, smaller volume, and more uniform tension insulators joints, Dr. Rong developed a petal overlap joint design, as shown in **Figure 10**.

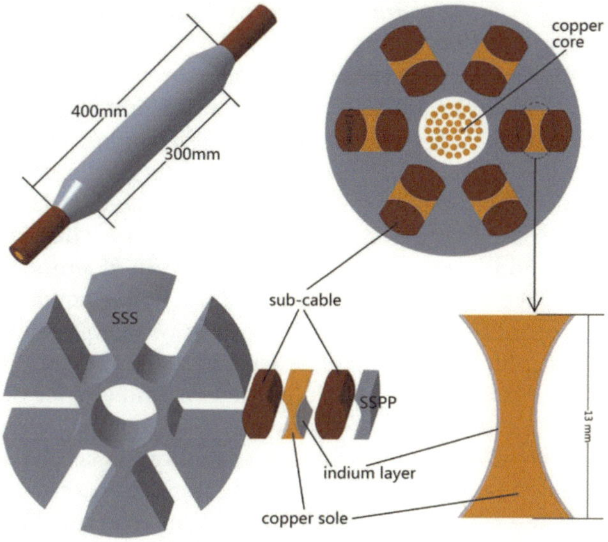

Figure 10.
The petal overlaps joint.

The assembly procedure of the petal overlap joint is set referring to twin-box joint:

a. Untwist the cable, get six petals and one copper core with length of 600 mm, and remove the nickel on the outer layer strands of petals by reverse electroplating process, but the inner layer remained.

b. Plate a 4 μm thickness silver plating-layer on sub-cables and copper soles, coat about 0.2 mm thickness pure indium layer on both sides of copper soles, and clean the surface of the indium layer.

c. Put one copper core through the central hole of stainless-steel support (SSS), and connect it with another copper core by copper strand welded.

d. Put six petals of one busbar, six copper soles, six petals of another busbar and stainless-steel pressure plates (SSPP) into the six grooves of SSS in turn, the length of SSS and copper is 300 mm.

e. Exert about 100 tons of pressure at room temperature (referring to twin-box assembly) to the SSPP till about 14 days later, and SSS and SSPPs are welded together.

In the petal overlap joint, the superfluid helium coolant flows through the bundle void of petals and core hole without external cooling pipes. It is known that cylinder makes insulators tension more uniform. In the complex electric-magnetic field, there will be AC losses and joule heating in the superconducting joint, which are important parameters, will be discussed in detail in the next subsection.

3.2 The joule heating and AC losses of petal overlap joint

Just like twin-box joint, in the resistance calculation model of petals overlap joint, **Figure 11(a)** shows the current flows direction, the current is carried by superconducting filaments only in the cable, while the copper matrix, indium layers, silver layers, and copper soles carry the current together in the joint, thus the joint resistance consists of copper matrix resistances, indium layer resistances, silver layer resistances, copper sole resistances, and contact resistances. The filament-to-matrix contact resistances could be estimated by the comparison of simulation with experimental results in the Twente Press Experiment [27–30]. As shown in **Figure 11(b)**, in the joint resistance calculation model, the two cables were equivalent to a homogeneous superconductor plus an equivalent resistance layer, the equivalent resistance layer represented the resistance of filament-to-matrix and copper matrix. Then the joint resistance consists of two equivalent layer resistances, two copper sole resistances, three indium layer resistances, and six silver layer resistances.

In the resistance simulation, because of low enough resistivity of 3.6×10^{-3} nΩm at 4.5 K temperature and thin enough thickness of 4 μm, the resistance contribution of silver is neglected [31], the superconductor resistivity is set at a low enough value of 1.0×10^{-6} nΩm to simulate zero resistance, as listed in **Table 3**. According to the experiment of joint assembly, the thickness of indium layer is set 0.2 mm, whose electrical conductivity is 7.46×10^{10} S/m at temperature of 4.5 K, according to Indium Corporation, [32, 33], as listed in **Table 3**. The current flows from cable to cooper sole through the outer layer strands, and the strand diameter 0.73 mm, then, the thickness of the equivalent resistance layer is set 0.365 mm, and the resistivity of the resistance layer is set equaling effective filament resistivity [27].

The twin-box resistance calculation model is used to verify the accuracy of the equivalent of the model, as shown in **Figure 12(a)**, in the joint resistance calculation model, a 68 kA DC load is applied on one busbar cable, and grounded another cable, the calculation resistance of the twin-box joint is 0.201 nΩ, while the test resistance was 0.2 nΩ carried out in ASIPP. Thus, the joint resistance calculation model was acceptable. In the petal overlap joint resistance calculation model, the petal overlap

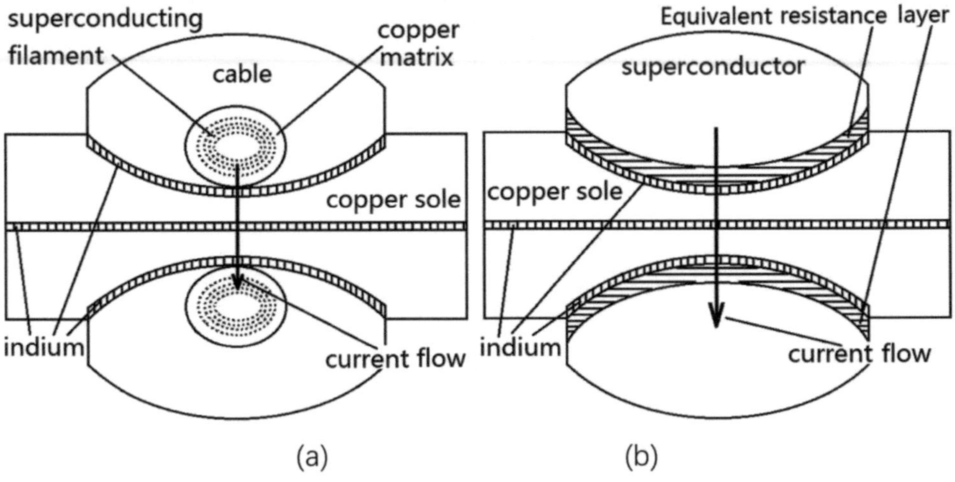

Figure 11.
The cross section and current flow diagrams of (a) realistic joint and (b) resistance calculation model.

material	Resistivity(nΩm)	Electrical conductivity (S/m)
Silver	3.6×10^{-3}	2.78×10^{11}
Superconductor	1.0×10^{-6}	1.0×10^{15}
Equivalent resistance layer	1.7	5.88×10^{8}
Copper	0.18	5.56×10^{9}
Indium	1.34×10^{-2}	7.46×10^{10}

Table 3.
The resistivity of the joint calculation model.

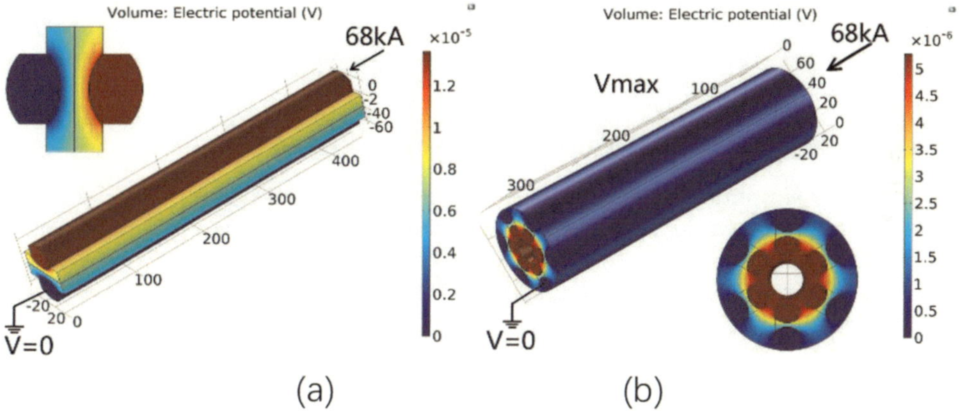

Figure 12.
The electric potential distribution in (a) twin-box joint and (b) petal overlap joint used for joint resistance calculation.

joint model was applied a 68 kA DC load on the petals of one busbar cable, and grounded the petals of another cable, as shown in **Figure 12(b)**. The petal overlap joint resistance was calculated at 7.76×10^{-2} nΩ.

In 15 MA reference scenario, the significant field variation only presents at the location near the terminal joint, therefore the joint AC losses are only significant in the terminal joints. AC losses of the joint were calculated with the following method [34]:

$$B + \tau \frac{dB}{dt} = B_a \; P_{AC-joint} = \frac{n\tau}{\mu_0} \left(\frac{dB}{dt} \right)^2 V, \tag{41}$$

Where B_a stands for the external magnetic field, whose direction is related to the studied contribution, while B stands for the local field. The demagnetization factor n is 1 for a slab, while 2 for a round. $P_{AC-joint}$ stands for AC losses of the twin-box joint, V is the effective shielded volume, and τ stands for the time constant of induced currents.

During the AC losses of the petal overlap joint calculation, the petal overlap joint is equivalent to six twin-box joints, while the AC losses of SSS and SSPP are neglected because of its high resistivity. **Figure 13** shows the geometrical simplification of one

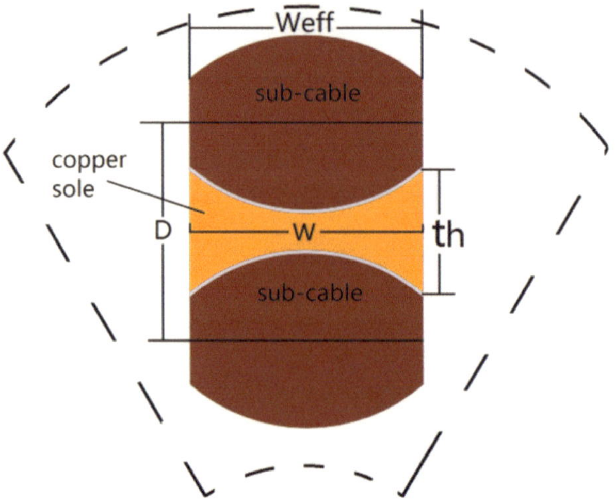

Figure 13.
The geometrical simplification of twin-box joint simplified by petal overlap joint.

twin-box joint simplified from petal overlap joint. Then the AC losses calculation method of twin-box could be used to calculate the AC losses of petal overlap joint:

a. The field normal to the copper sole causes coupling loss and eddy current loss, whose coupling time constant τ_{trans} carried out by the following [35]:

$$\tau_{trans} = \frac{\mu_0 W^2}{12\rho_{copper}}, \tag{42}$$

As shown in **Figure 13**, and listed in **Table 4**, W stands for the width of the copper sole, whose resistivity ρ_{copper} is listed in **Table 3**. The effective volume (V_{trans}) equals the volume of the copper sole plus cables.

b. The field tangent to the copper sole and normal to the cables causes coupling loss and eddy current, whose coupling time constant τ_{cond} is carried out by the following [35]:

$$\tau_{cond} = \frac{L_1 - L_2 - 2M}{R_{m1} - R_{m2}}, \tag{43}$$

where L_1, L_2 stand for self-inductance of two cables in the joint, while M stands for the mutual-inductance of the two cables. R_{m1} and R_{m2} stand for equivalent resistance of the two cables [35]. The effective volume (V_{cond}) equals the volume of half-cables.

c. The field produces the inter-cable current, which causes loss referring to the copper sole and normal conductor of the cables. The coupling time constant τ_{joint} could be carried out by the following [35]:

$$\tau_{joint} = \frac{\mu_0 LD}{12W_{eff}R_{joint}}, \tag{44}$$

Where L, W_{eff}, and D, stand for the joint length, the effective width of sub-cables of petal overlap joint, and the distance of two sub-cables in a petal overlap joint respectively, listed in **Table 4**. R_{joint} stands for the resistance of one simplified twin-box joint. The effective volume (V_{joint}) is carried out by the following:

$$V_{joint} = L \bullet W_{eff} D. \tag{45}$$

The AC losses of the three contributions are carried out by Eq. (47) with field variations referring to IDM_45SA69 of ITER organization, as listed in **Table 5**. The AC loss of one twin-box joint simplified from the petal overlap joint is about 0.197 W, as listed in **Table 5**. Then the AC loss of total petal overlap joint is 1.182 W.

Ohmic heating produced by the joint resistance and transport current was calculated by the following equation:

$$P_{ohmic} = R_{joint} I_{op}^2, \tag{46}$$

Geometrical parameters	Units	One simplified twin-box joint in petal overlap joint
W_{eff}	mm	13
W	mm	13
th	mm	7.2
D	mm	12
L	mm	300
V_{trans}	mm^3	0.076
V_{cond}	mm^3	0.061
V_{joint}	mm^3	0.047

Table 4.
The geometrical parameters of petal overlap joint.

Calculation losses	Units	One simplified twin-box joint in petal overlap joint
τ_{trans}	s	0.098
τ_{cond}	s	0.2
τ_{joint}	s	20.49
P_{trans} [a]	nW	7.76×10^{-3}
P_{cond} [b]	W	0.18
P_{joint} [c]	W	0.017
$P_{AC-joint}$	W	1.182
P_{ohmic} [d]	W	0.36
P_{total}	W	1.542

Table 5.
The losses of and petal overlap joint.

where R_{joint} is joint resistance, and I_{op} is the transport current. The Ohmic heating is calculated at 0.36 W produced by petal overlap joint, as listed in **Table 5**.

The AC losses and joule heating could cause a temperature rise in the main busbar, which could reduce the temperature margin. Then, the temperature rise caused by petals overlap superconducting joints will be analyzed carefully in the next subsection.

3.3 The temperature rise caused by the petal overlap joint

According to ITER requirements, the design temperature margin of the main busbar should be above 2.5 K. The temperature margin (ΔT) reflects the ability of removing heat of surrounding helium before the quench of the superconducting strands [36, 37].

The THEA code is used to simulate the main busbar with petal overlap joints in 15 MA scenario. During operation, there are AC losses of busbar and joints, joule heating of the joints, heat exchange in the joints, and the heat load from supports and HTS (high-temperature superconductor), which cause a temperature rise, and are taken account into the THEA model. Busbar cable AC losses (including coupling and hysteresis losses) have been calculated by M. Nannini (ITER organization IDM_45SA69), whose maximum value is about 1.33 W/m. In the THEA model, busbar cable AC losses are assumed constant along the busbar, which equals the maximum value. There are five joints in each main busbar at the locations of 0 m, 6.5 m, 12.9 m, 25.3 m, and 36 m, respectively.

As shown in **Figure 14(a)**, there are five temperature rises in the location of the joints, the temperature increases along the main cable same direction with the helium flowing, and the maximum temperature in the busbar cable with petal overlap joints was 5.5 K, which occurred at the location of the fifth joint, this maximum temperature was added by HTS current leader heating mainly.

The current sharing temperature (Tcs) refers to the able pattern and the operation current, and its distributions of respective busbar did not change with time, as shown in **Figure 14(b)**, the current sharing temperatures were almost

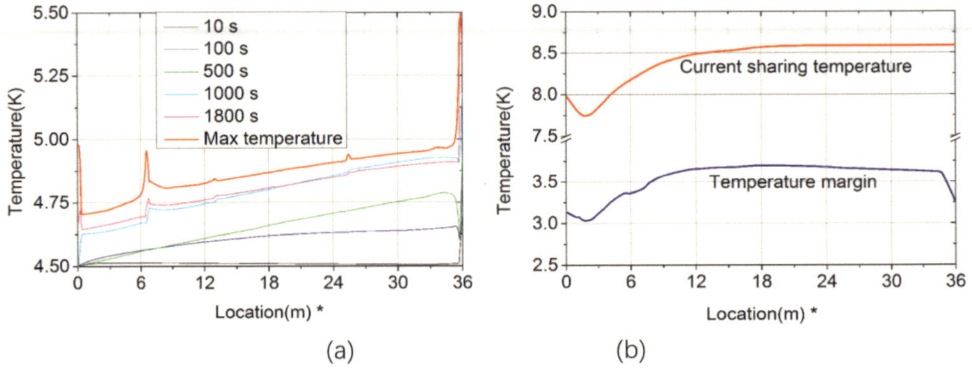

Figure 14.
*(a) The temperature distribution of busbar with petal overlap joints, and (b) current sharing temperature and temperature margin distributions of busbar with petal overlap joints (*the coordinate of "location(m)" in the figures is the location of main busbar, and 0 m is located in the end of busbar near the magnet).*

impervious to joint heating. The temperature margin was calculated by the following equation:

$$\Delta T = T_{cs} - T_{max}, \tag{47}$$

where T_{max} is the maximum temperature at each location (as shown in **Figure 14(b)** thick red lines) of the busbar cable. The lowest temperature margin of cable with petal overlap joints was 2.85 K, which reached ITER requirements.

The effects of joints to the main busbar are analyzed carefully; however, there are disturbances in the main busbar conductor, which could cause a nonrecoverable quench. In the next section, the stability and quench behavior of the main busbar will be discussed in detail.

4. Stability and quench behavior analysis of the ITER TF main busbar

In the ITER TF main busbar, there are static and AC losses in the superconducting conductor, such as heat exchange in joint, joule heating in joint, AC losses in joint, control losses, radiation from 80 K cooling, vacuum barrier conduction, external support and peak AC losses in conductor, which could cause a temperature rise in the conductor. The temperature rise could reduce the temperature margin of the superconducting conductor, and the temperature margin should be set large enough to guarantee the stability of the conductor after covering the static and AC losses. During superconducting magnet operation, the superconducting conductor is considered to quench, once the operation temperature exceeds the current sharing temperature.

During superconducting magnets operation, the mechanical displacement, increased local stress and local shear force could cause the local thermal deposition in the superconductor, which could cause an unrecovery quench in the superconducting conductor. The smallest energy that causes an unrecovery quench is called minimum quench energy (MQE).

After quench, the current is carried by the normal conductor, and the temperature rises, then the current is cut down, and then the temperature of the conductor rises unceasingly because of the residual energy left in the magnet. The maximum temperature is called hotspot temperature, which refers to the ratio of superconductor and normal conductor.

In this section, the temperature margin, minimum quench energy, and hotspot temperature of the ITER TF main busbar will be discussed in detail.

4.1 The temperature margin of the ITER TF main busbar

The TF main busbar has a copper core (C0), as shown in **Figure 15**. which is cooled internally with forced flow supercritical helium at an inlet temperature of 4.5 K and inlet pressure of 6.0 bar, and the operation current is 68 kA [2, 9, 10, 25], as listed in **Table 1**, the bundle helium flows in the bundle void, while hole helium flows in the rope void, shown in **Figure 15**.

The temperature margin provides a margin against uncertainties in the strand performance (e.g., critical current, temperature) and a capacity for heat absorption in

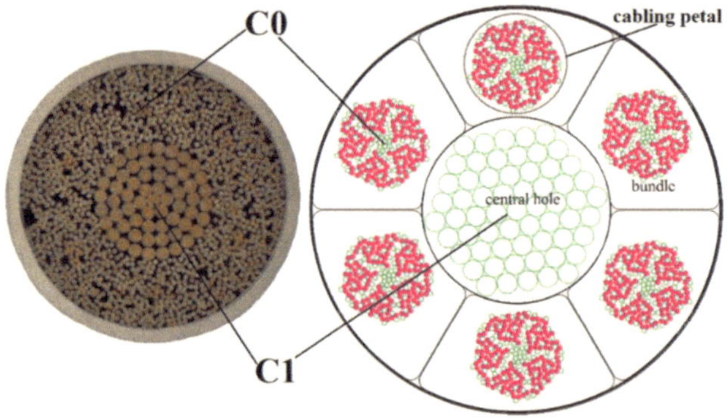

Figure 15.
Schematic of the main busbar cross-sectional layout.

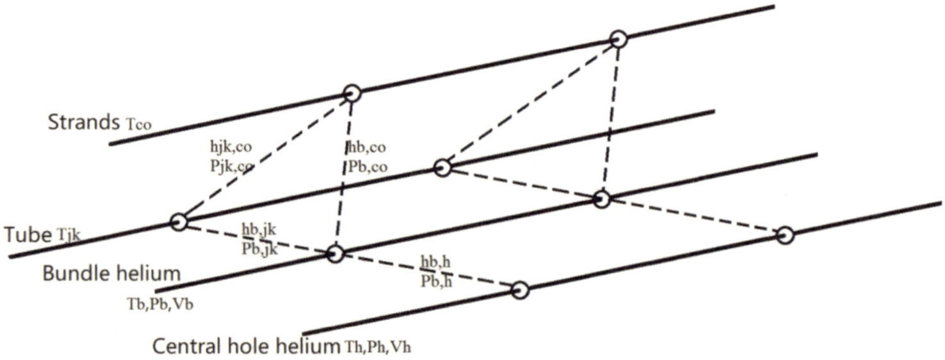

Figure 16.
Schematic of the one-dimensional finite element model of TF main busbar.

the surrounding helium before the strands quench [36, 37], the temperature margin could be carried out by Eq. (47).

The temperature behavior of the busbar during 15 MA scenario is analyzed by one-dimensional finite element numerical simulation. The following simplifying assumptions of Gandalf are significant for this study, as shown in **Figure 16**:

a. the current is uniformly distributed among the strands;

b. the disturbance is applied uniformly in the conductor cross section;

c. all copper in the cable is included as a stabilizer;

d. the two-fluid model (forced flow helium flowing in the hole and the bundle) is used in this analysis.

As shown in **Figure 16**, the coupling of the model is given by heating exchanging between the four sub-one-dimensional models, and the liquid helium fluid abides by the law of conservation of mass (the continuity equation), the law of conservation of energy (Bernoulli's equation), and the law of conservation of momentum (momentum equation) [38–40]:

$$\begin{aligned}
&\frac{\partial V_b}{\partial t}+V_b\frac{\partial V_b}{\partial x}+\frac{1}{\rho}\frac{\partial P}{\partial x}=-2f_b\frac{V_b|V_b|}{D_{hb}}\\
&\frac{\partial V_h}{\partial t}+V_h\frac{\partial V_h}{\partial x}+\frac{1}{\rho}\frac{\partial P}{\partial x}=-2f_h\frac{V_h|V_h|}{D_{hh}}\\
&\frac{\partial P}{\partial t}+a_b\rho c^2\frac{\partial V_b}{\partial x}+a_h\rho c^2\frac{\partial V_h}{\partial x}+V\frac{\partial P}{\partial x}=\Phi(a_bV_bF_b+a_hV_hF_h)+\frac{\Phi Q_{He}}{A_{He}}\\
&\frac{\partial T_{He}}{\partial t}+a_b\Phi T_{He}\frac{\partial V_b}{\partial x}+a_h\Phi T_{He}\frac{\partial V_h}{\partial x}+V\frac{\partial T_{He}}{\partial x}=\frac{a_bV_bF_b+a_hV_hF_h}{C_v}+\frac{Q_{He}}{A_{He}\rho C_v}\\
&AC\frac{\partial T_{co}}{\partial t}-A\frac{\partial}{\partial x}\left(k\frac{\partial T_{co}}{\partial x}\right)=-Q_{He,co}+Q_{joule}+Q_{ext}-Q_{co,jk}\\
&A_{jk}C_{jk}\frac{\partial T_{jk}}{\partial t}-A\frac{\partial}{\partial x}\left(k_{jk}\frac{\partial T_{jk}}{\partial x}\right)=-Q_{He,jk}+Q_{joule,jk}+Q_{ext,jk}-Q_{co,jk}
\end{aligned}, \quad (48)$$

Where V_b and V_h are the flow velocity of bundle and central helium respectively, D_{hb} is the hydraulic diameter of the bundle helium, P is the pressure of helium, c is the acoustic velocity of super-critical helium, Φ is the Grüneisen parameter of helium, ρ is the density of helium, C_v is the specific heat at constant volume of helium, A, A_{jk}, A_b, A_h, and A_{He} are the cross-sectional areas of conductor, jacket, bundle helium, central hole helium, and the total helium, respectively, and V, a_b, a_h, F_b, and F_h could be given by the following:

$$\begin{gathered}
V=a_bV_b+a_hV_h\\
a_b=\frac{A_b}{A_{He}}\quad a_h=\frac{A_h}{A_{He}}\\
F_b=2f_b\frac{V_b|V_b|}{D_{hb}}\quad F_h=2f_h\frac{V_h|V_h|}{D_{hh}}
\end{gathered}, \quad (49)$$

Where f_b and f_b are the coefficients of friction of bundle and central helium, respectively.

The equation could be solved by the finite element method, time is dispersed by the numerical approach method, while the spatial is dispersed by the Galerkin method, and the upwind is used to simulate the flow direction of helium. In this study, the Gandalf code is used to study the behavior of the TF main busbar in 15MA plasma current reference scenario.

The temperature margin is simulated by THEA code. In the 15 MA reference scenario, there AC losses, joule heating, and other losses in the main busbar, which could disturb the stability of the busbar, and the superconducting cable must be able to absorb them without causing a quench. In the THEA model, all losses are applied directly into the conductor as the external linear heating input, which is a square wave in space and time, which is assumed uniform along the conductor, as shown in **Table 6**. In the 15 MA scenario, AC losses are estimated and tested in SULTAN [15, 20, 41], which is 1.38 W/m in the plasma building phase (0–200 s), and is neglected after 200 s. Then the external heating input of the model is shown in **Figure 17**, and the following conditions are used for the model [2, 9, 10, 25]:

Parameters	Value	Units
Heat exchange in joint	9.3	W
Joule heating in joints	47	W
AC losses in joint	1.4	W
Control losses	3.3	W/m
Radiation from 80 K	3.3	W/m
Vacuum barrier conduction	2.73	W
External support	7.6	W
Peak AC loss in cable	1.38	W/m

Table 6.
Summary of the static and AC losses for the main busbar in 15 MA scenario.

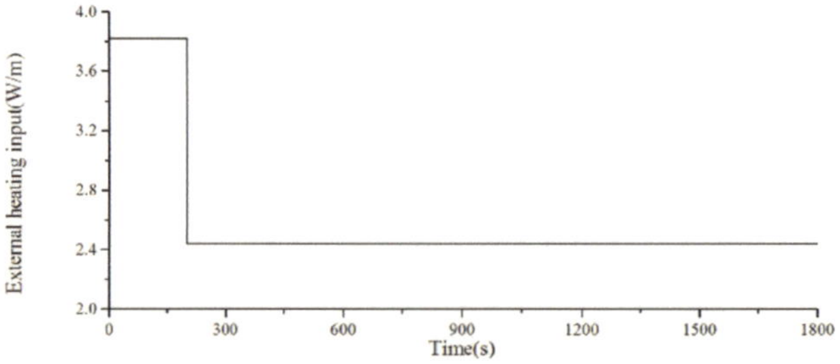

Figure 17.
External heating input of Gandalf model during 15 MA scenario.

a. The cable length is set 30 m for the worst case and without jumpers, the inlet temperature is 4.5 K, inlet pressure is 6.0 bar, and the mass flow of helium is 10.5 g/s.

b. The time step is set as 1 s, and the maximum mesh sizes is 0.1 m for a stable purpose.

c. The cross-sectional areas of bundle helium and the hole helium are 417.8 mm^2 and 47.9 mm^2 repectively.

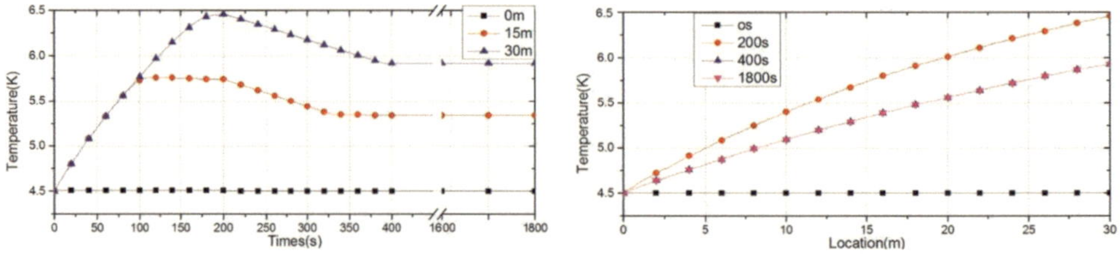

Figure 18.
Temperature behavior of the TF main busbar conductor in 15 MA reference scenario.

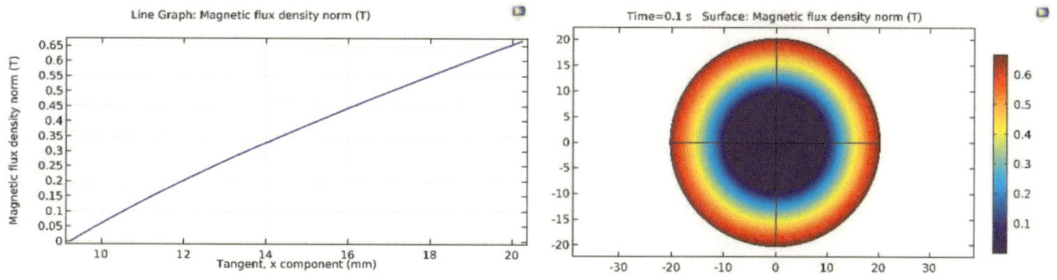

Figure 19.
Distribution of self-field in the TF main busbar cross section.

d. The external heat flux input (all losses included) is uniform at 3.82 W/m from 0 to 200 s, and 2.44 W/m (ac losses excluded) for the remaining times.

According to the model, in the worst case, the maximum temperature is 6.45 K at the time of 200 s and the location of 30 m in the 15 MA scenario, as shown in **Figure 18**.

Temperature rise could cause a current sharing in the matrix, then there is electric field in the superconducting cable. In the engineering application, a parameter named the current sharing temperature is defined, the current sharing temperature equals the temperature, which causes an electric field value of 10 μV/m. In the 15 MA scenario with 3.0 T field, 4.5 K temperature and 68 kA operation current, the current sharing temperature of the main busbar have been carried out by Gandalf code, which equals 7.62 K. In the Gandalf model, the magnetic field distribution in the cross section of the cable is assumed uniform; however, the field is not uniform in the cross section of the cable actually, then the current sharing temperature should be estimated further exactly.

In 15 MA scenario, the self-field distribution in the main busbar cross section is simulated by the finite element method in the COMSOL model, which is not uniform, as shown in **Figure 18**.

In this study, the current sharing temperature is estimated by E-J characteristic. As shown in **Figure 19**, the self-field distribution is axis-symmetrical, then the bundle zone is divided into 220 annuluses, the difference between two adjacent annuluses is 0.05 mm. Thus, the cross-sectional area of the *ith* annulus is:

$$A_{sci} = \frac{A_{sc}}{\pi\left(r_{out}^2 - r^2\right)}\pi\left(\left(x_i + dx\right)^2 - x_i^2\right). \tag{50}$$

Then the E-J characteristic could be turned into the following:

$$E_{av} = \frac{E_c}{A_{sc}}\sum_i^N A_{sci}\left(\frac{J_{op}}{J_c(B_i, T)}\right)^{n(J_c)}, \tag{51}$$

where B_i is the magnetic flux density of the *ith* annulus.

The current sharing temperature of full-size main busbar sample is tested in SULTAN, the temperature is 4.22 K, the background field is 3.22 T, and the operation current is 45.5 K. the tested current sharing temperature is 6.98 K, and the current sharing temperature from Gandalf model is 7.34 K, while current sharing temperature calculated by Eq. (51) is 7.09 K. This indicates that the method is much more accurate than the simulated value by the Gandalf model.

Operation current	kA	68
Background field	T	1.89
T_{cs} Gandalf model	K	7.62
T_{cs} Computed from COMSOL model	K	7.37

Table 7.
Current sharing temperature of main busbar used in 15 MA scenario.

In 15 MA scenario, the operation current of the TF main busbar is 68 kA current, and the background field equals 1.89 T. Then the current sharing temperature of the main busbar could be carried out by solving Eq. (51), which is 7.37 K, as shown in **Table 7**.

In the 15 MA scenario, the current sharing temperature is estimated by Gandalf model, which is 7.62 K, and that is estimated by Eq. (51) at 7.37 K. Then the temperature margin is estimated by Gandalf code, which is 3.12 K, and that is estimated by Eqs. (51) and (47) at 2.87 K.

4.2 The minimum quench energy (MQE) of TF main busbar in 15 MA scenario

Transient heating disturbance may cause a quench in the superconducting cable, then for reflecting the disturbance, a parameter called minimum quench energy (MQE) is design. The MQE is defined as the minimum trial energy, which causes a nonrecoverable quench. Generally, during MQE estimation, the energy pulse is a small spatial extent (1–10 mm) and short duration (40–100 μs) applied to the conductor, which is just sufficient to initiate a quench [42, 43].

The MQE of the TF main busbar is carried out by the one-dimensional finite element method in the Gandalf model, the temperature in the cross section of the conductor is assumed uniform, then a one-dimensional equation of heat conduction of the TF main bus bar is carried out [44–46]:

$$A\frac{\partial}{\partial x}\left(k(T)\frac{\partial T}{\partial x}\right)+I_{op}^{2}R(T)+Q_{ext}(x,t)-h(T)P_{wet}(T-T_{He})-AC(T)\frac{\partial T}{\partial t}=0,\quad (52)$$

Where $k(T)$ is the thermal conductivity of the conductor (W/(mk)), T is the temperature of the conductor (K), T_{He} is the temperature of helium (K), I_{op} is the operation current (A), $R(T)$ is effective resistance per unit length (Ω/m), A is the area of superconductor and matrix (m^2), Q_{ext} is the external thermal deposition (W/m), $h(T)$ is the thermal conductivity from conductor to helium (W/(m^2K)), P_{wet} is the wet perimeter of the conductor, $C(T)$ is the specific heat per unit volume (J/(m^3K)), and t is the time.

In this study, the heating pulse is the linear input applied in the conductor, which have spatial extents of 1, 3, 6, 8, and 10 mm and durations of 40, 60, 80, and 100 μs. The MQE of the main busbar is estimated by Gandalf model, and the following shows the input parameters of the model:

a. The initial external heating is applied in the middle of the model, and the time extent is from 0 to heating pulse duration [45], as shown in **Figure 20**.

a. The time step is set as 10^{-6}–10^{-2} s, and the mesh size in the heating pulse region is set as 0.01 mm to keep a stable numerical analysis.

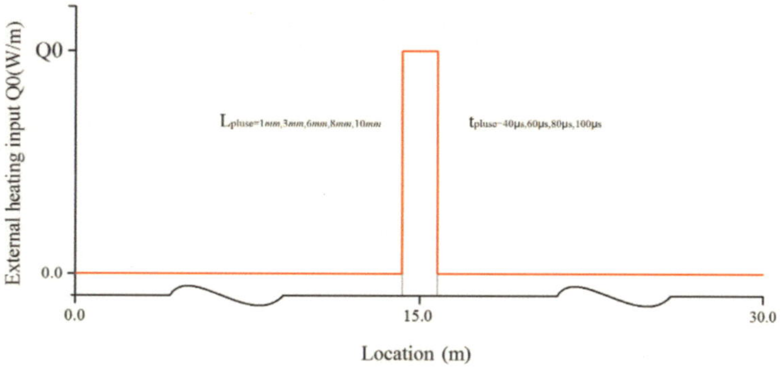

Figure 20.
External heating pulse input Q_{ext} in the Gandalf model.

Heat energy (J)	Heat flux (MW/m)	Recovery time (ms)
1.43	35.8	69.0
1.44	36.0	76.0
1.46	36.4	Never recovery
1.47	36.8	Never recovery

Table 8.
MQE simulated by Gandalf model with heating pulse input region length of 1.0 mm and duration of 40 μs.

b. The total delay time equals the detective time plus the delay time of the main busbar quench protection system, which equals 3.0 s, while the discharge time constant equals 11.0 s.

In the condition of heating input spatial extent of 1.0 mm and duration of 40 μs, the different heating pulse input value causes different quench behavior, as listed in **Table 8**. The MQE of the TF main busbar is set as the heating input value as following, which is 1.46 J:

$$\frac{MQE - Q_{recovery}}{MQE} < 1\%,$$

Where $Q_{recovery}$ cause a recoverable quench, and MQE causes a nonrecoverable quench.

During the MQE heating input, an instantaneous propagation quench presents, which gives a normal length L_{nor} of 26 cm. Shortly after the end of the heating pulse duration of 40 μs, the heating pulse becomes 0, the quench propagation stops, and the quench began recovery due to helium coolant heating removing. However, the normal length just becomes smaller but not vanish ($L_{nor} = 10.3cm$ at t = 20 ms), and there is joule heating produced in the normal cable. At this time, the recovery is only partial, and the quench will propagate unceasingly, because heating production is higher than heating removing. At the time of 23 ms, the normal length L_{nor} shrinks again, and at the time of 103 ms, the joule heating beats the enhanced cooling, and drives the conductor toward an irreversible quench, as shown in **Figure 21** red solid lines.

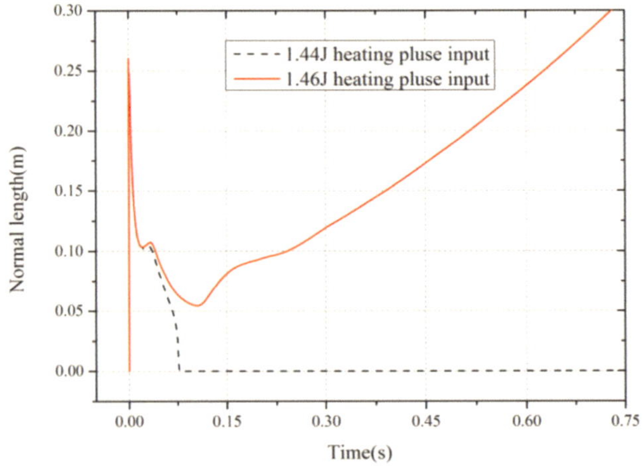

Figure 21.
Normal length in the case of quench triggered by 1.46 and 1.44 J heating pulse input.

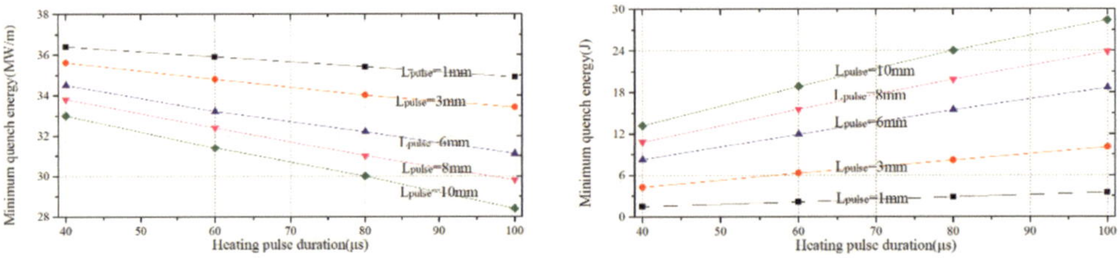

Figure 22.
MQE of the TF main busbars with heating pulse input spatial extents of 1, 3, 6, 8, and 10 mm and durations of 40, 60, 80, and 100 μs.

During the 1.44 J heating input, the quench propagation behavior is the same with the MQE heating input till the time of 33 ms; however, at this time the enhanced helium cooling is enough to drive a recovery in a few milliseconds, and then the full recovery is got at the time of about 76 ms, as shown in **Figure 21** dash lines.

The MQEs of the main busbar with different heating input spatial extents and durations are carried out by the Gandalf model, and **Figure 22** shows the MQEs of the main busbar with the heating input spatial the of 1, 3, 6, 8, and 10 mm and durations of 40, 60, 80, and 100 μs. The lower heating input per length is needed to cause a nonrecoverable quench, the larger heating pulse extents and durations.

The MQE could be affected by the operation current, because smaller current produces lower joule heating, then larger MQE is needed to cause a nonrecoverable quench. With the heat pulse spatial extent of 1.0 mm and duration of 40 μs, MQEs with different operation current are carried out by Gandalf code, as shown in **Figure 23**. It is clear that the higher operation current, the smaller MQE is needed to trigger a nonrecoverable quench.

4.3 The hotspot temperature of TF main busbar in 15 MA plasma current scenario

After an nonrecoverable quench caused by MQE occurs in the TF main busbar, the quench is detected after the delay time and during detective time, then the current of

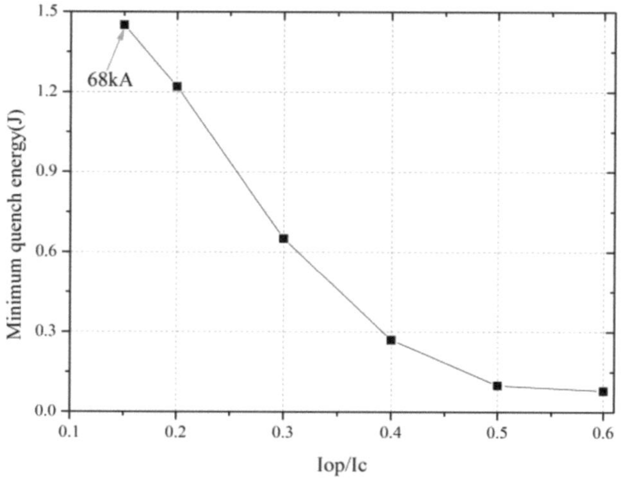

Figure 23.
MQE of the TF main busbars with different currents.

the magnet is cut down, the quench propagates along the conductor and the conductor temperature rises, till the current approaches zero gradually, in which the maximum temperature is called hotspot temperature. The hotspot temperature is one of key parameters, which refers to the copper strands ratio in the conductor, the exorbitant hotspot temperature could cause damage to the conductor. The highest hotspot temperature allowable is 150 K according to the ITER design criterion [47].

The MQE is evaluated by one-dimensional Gandalf code, the quench propagation behavior of the TF main busbar is simulated in the un-adiabatic condition. The conductor quench is initiated at the center of the conductor. The following conditions are used for the numerical simulation:

a. The inlet temperature is 4.5 K, the inlet pressure is 6.0 bar, and the mass flow is 10.5 g/s [47, 48].

b. The time step and the mesh size are set as 1–10 ms and 1 mm at the location of heating input to keep a stable numerical analysis, and 0.1 m for the rest.

c. MQE with the extent of 60 mm and duration of 0.5 ms is applied in the middle of the model.

d. The delay time equals 3 s, which is the delay time of 1 s plus the detective time of 2 s, and discharge time is 11 s [47].

In the Gandalf model, the copper strands and the copper matrix of the superconducting strands is not separated, and the Gandalf model is a two-fluid model including central hole helium and bundle helium, which is acceptable according to experimental results [49]. The MQE is applied in the middle of the model, compared with the joule heat, the MQE is low enough to be neglected during hotspot temperature estimation. The maximum temperature in the total discharge cycle is about 57.1 K with 100% extra copper, which is lower than the ITER requirement value of 150 K. In the condition of adiabatic, there is only heating conducting of conductor to remove

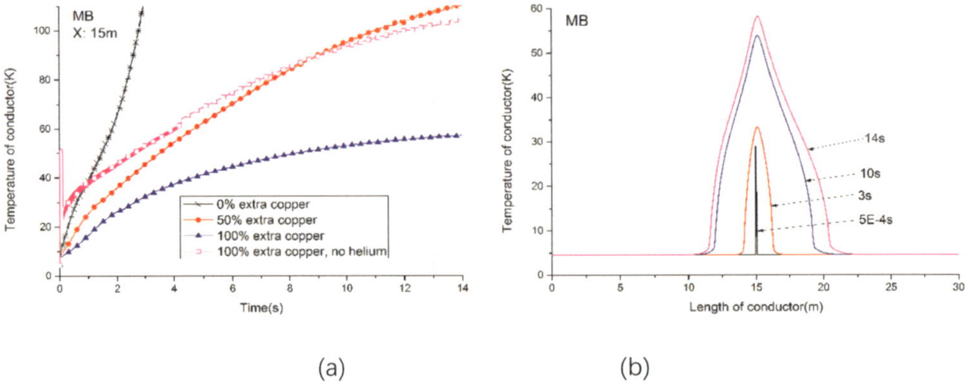

Figure 24.
(a) Temperature of conductor after quench in different extra copper content and in the case of no helium, (b) temperature distributions after quench in MB.

heating, the maximum temperature is 104 K at one discharge cycle, which is lower than the ITER requirement value of 250 K in in adiabatic condition, as shown in **Figure 24(a)**.

At the end of the heating pulse duration, the maximum temperature reaches 28 K at the middle of the model, in which the heating pulse is applied, and a nonrecoverable quench is, as shown in **Figure 24(a)**. Because of the quench, superconductor becomes normal conductor, the current is carried by copper conductor and normal superconductor jointly, which produces the joule heating, the quench propagates, and the temperature of the conductor rises gradually. The quench presents firstly at the location of the external heating pulse, and then the quench propagates along the conductor in both sides because of joule heating, the helium flows from 0 m to 30 m, so the temperature curve along the conductor presents as a dissymmetry parabola, as shown in **Figure 24(b)**.

5. Conclusions

In this chapter, AC losses calculation of ITER TF main busbar was discussed, the hysteresis loss refers to superconducting filaments, while the coupling losses refer to the pattern of the conductor, and a developed method of contact area ratio calculation is introduced. For a small impact on the main busbar form joints, a development of petal overlap joint is discussed in detail, and its AC losses and joule heating are calculated carefully; furthermore, the temperature rise caused by the joints is analyzed carefully.

The characteristics of cooling helium, inlet temperature, operation current, and external copper are some important parameters in the main busbar design, which could affect the stability and quench behavior of the main busbar. The current sharing temperature refers to the operation current, is regarded as the criterion of quench, which is calculated taking self-field account into. The minimum quench energy referring to inlet temperature and cooling helium is simulated carefully, while the hotspot temperature referring to external copper is simulated. This chapter hopes to provide some help for superconducting magnets research.

Author details

Jian Rong
School of Nuclear Science and Technology, Lanzhou University, China

*Address all correspondence to: jianrong@mail.ustc.edu.cn

References

[1] Song Y, Bauer P, Bi Y, et al. Design of the ITER TF magnet feeder systems. IEEE Transactions on Applied Superconductivity. 2010;**20**(3):1710-1713

[2] Fu, Peng, Song Y, Wu Y, et al. Recent Progress of ITER Package in ASIPP. 2018

[3] Jin F. The Theoretical and Experimental Research of Stability of CICC of HT-7U [Doctoral dissertation]. Hefei: Institute of Plasma Physics Chinese Academy of Sciences; 2002

[4] Bean CP. Magnetization of hard superconductors. Physical Review Letters. 1962;**8**(6):250

[5] Prigozhin L. The Bean model in superconductivity: Variational formulation and numerical solution. Journal of Computational Physics. 1996; **129**(1):190-200

[6] Jian R. The Electromagnetics Key Problems of TOKAMAK Superconducting Magnet and Feeder Systems [Doctoral dissertation]. Hefei: University of Science and Technology of China; 2016

[7] Bottura L. A practical fit for the critical surface of NbTi. IEEE Transactions on Applied Superconductivity. 2000;**10**(1): 1054-1057

[8] Seeber B. Handbook of Applied Superconductivity. Bristol: Institute of Physics Pub.; 1998

[9] Qin J, Wu Y, Yu M, et al. Manufacture of ITER feeder sample conductors. Fusion Engineering and Design. 2013;**88**(9–10):1461-1464

[10] Liu B, Wu Y, Qin J, et al. The performance test and analysis of ITER Main and correction Busbar conductor. Fusion Engineering and Design. 2013;**88**: 2802-2809

[11] ITER Organization Magnet Team, ITER Magnet Feeder Busbar Design Study, ITER_D_ 2ZUR4A. 2010

[12] Miao H, Huang Y, Meinesz M, et al. Development of Bi-2212 round wires for high field magnet applications. AIP Conference Proceedings. American Institute of Physics. 2012;**1435**(1):315-324

[13] Zhou C, Miyoshi Y, Van Lanen E, et al. Inter-filament resistance, effective transverse resistivity and coupling loss in superconducting multi-filamentary NbTi and Nb3Sn strands. Superconductor Science and Technology. 2012;**25**(6): 065018

[14] Qin J, Wu Y. A 3D numerical model study for superconducting cable pattern. Fusion Engineering and Design. 2010; **85**(1):109-114

[15] Rong J, Huang X, et al. AC loss of ITER feeder Busbar in 15MA plasma current reference scenario. Journal of Fusion Energy. 2016;**35**:173-179

[16] Ilyin YA, Nijhuis A, Abbas W, et al. Effect of cyclic loading and conductor layout on contact resistance of full-size ITER PFCI conductors. IEEE Transactions on Applied Superconductivity. 2005;**15**(2):1359-1362

[17] Huang J, Ilyin Y, Wessel WAJ, et al. Contact resistance, coupling and hysteresis loss measurements of ITER poloidal field joint in parallel applied magnetic field. Superconductor Science and Technology. 2022;**35**(2):025016

[18] Yagotintsev KA, Wessel WAJ, Vostner A, et al. Overview of verification

tests on AC loss, contact resistance and mechanical properties of ITER conductors with transverse loading up to 30 000 cycles. Superconductor Science and Technology. 2019;**32**(10): 105015

[19] Ciazynski TS. A model for calculating a.c. losses in multistage superconducting cables. Cryogenics. 1996;**36**:1039-1049

[20] Yagotintsev KA et al. Superconductor science and technology accepted manuscript overview of verification tests on AC loss, contact resistance and mechanical properties of ITER conductors with transverse loading up to 30,000 cycles. Superconductor Science and Technology. 2019;**32**:10

[21] Yoshida K, Takahashi Y, Isono T, et al. Updating the design of the feeder components for the ITER magnet system. Fusion Engineering and Design. 2005;**75**:241-247

[22] Henry D, Chalifour M, Forgeas A, et al. Process flow and functional analysis of the ITER cryogenic system. AIP Conference Proceedings. American Institute of Physics. 2010;**1218**(1): 676-683

[23] Sahu AK, Gung CY, Lu K, et al. Cryogenic engineering design of the ITER superconducting magnet feeders. IEEE Transactions on Applied Superconductivity. 2012;**22**(3): 4800604-4800604

[24] Devred A, Backbier I, Bessette D, et al. Status of ITER conductor development and production. IEEE Transactions on Applied Superconductivity. 2012;**22**(3): 4804909-4804909

[25] Calvi M, Bauer P, Bessette D, et al. Design proposal for the ITER feeder busbars. IEEE Transactions on Applied Superconductivity. 2010;**20**(3): 407-410

[26] Pong I, Vostner A, Bordini B, et al. Current sharing temperature of NbTi SULTAN samples compared to prediction using a single pinning mechanism parametrization for NbTi strand. Superconductor Science and Technology. 2012;**25**(5):054011

[27] Zhou C, Miyoshi Y, Van Lanen EPA, et al. Direct measurement of inter-filament resistance in various multi-filamentary superconducting NbTi and Nb3Sn strands. Superconductor Science and Technology. 2011;**25**(1): 015013

[28] Abbas W, Nijhuis A, Ilyin Y, et al. A fully automatic Press for Mechanical and Electrical Testing of full-size ITER conductors under transverse cyclic load. AIP Conference Proceedings. American Institute of Physics. 2004; **711**(1):51-58

[29] Rolando G. Cable-in-conduit superconductors for fusion magnets [PhD Thesis]. University of Twente; 2013

[30] Nijhuis A, ten Kate HHJ, Duchateau JL, et al. Control of contact resistance by strand surface coating in 36-strand NbTi CICCs. Cryogenics. 2001;**41**(1):1-7

[31] Khoshnevisan M, Pratt WP Jr, Schroeder PA, et al. Low-temperature resistivity of silver. Journal of Physics F: Metal Physics. 1979;**9**(1):L1

[32] Martin DL. Low-temperature electrical resistivity and superconducting transition temperature of indium-manganese alloys. Physical Review. 1965;**138**(2A):A464

[33] I. Corporation. Physical Constants of Pure Indium. Available from: http://www.indium.com/metals/indium/physical-constants. [Accessed: December 12, 2017]

[34] Ciazynski D, Martinez A. Electrical and thermal designs and analyses of joints for the ITER PF coils. IEEE Transactions on Applied Superconductivity. 2002;**12**(1):538-542

[35] Zani L, Ciazynski D, Heller R, et al. Study of current distribution in ITER TFMC NbTi Busbar III. IEEE Transactions on Applied Superconductivity. 2006;**16**(2): 864-867

[36] Zanino R, Bessette D, Richard LS. Quench analysis of an ITER TF coil. Fusion Engineering and Design. 2010; **85**(5):752-760

[37] Li S, Wu Y, Liu B, et al. Current sharing temperature test and simulation with GANDALF code for ITER PF2 conductor sample. Plasma Science and Technology. 2011;**13**(5):627

[38] Bottura L, Rosso C. Finite element simulation of steady state and transient forced convection in superfluid helium. International Journal for Numerical Methods in Fluids. 1999;**30**(8): 1091-1108

[39] Wang Q, Kim K, Yoon CS. Numerical model for thermal hydraulic analysis in cable-in-conduit-conductors. KSME international journal. 2000;**14**(9): 985-996

[40] Huang H, Wang Q. Heat Transfer Finite Element Analysis. Beijing: China Science publishing & Media Ltd.; 2011

[41] Miyoshi Y, Wessel WAJ, Krooshoop JG, et al. AC Loss Measurement with Transverse Load Cycles in Twente Press on Hitachi-81JNC002-S4T1. Intermediate Report UT-IO-2011-516. University of Twente; 2011

[42] Marinucci C, Savoldi L, Zannino R. Stability analysis of the ITER TF and CS conductors using the code Gandalf. IEEE Transactions on Applied Superconductivity. 1999;**9**(2):612-615

[43] Fu M, Xu X, Jiao Z, et al. Minimum quench energy and normal zone propagation velocity in MgB2 superconducting tape. Physica C: Superconductivity. 2004;**402**(3): 234-238

[44] Wilson MN, Wolf R. Calculation of minimum quench energies in Rutherford cables. IEEE Transactions on Applied Superconductivity. 1997;**7**(2): 950-953

[45] Wilson MN. Computation of Minimum Quench Energies in LHC Cables. Geneva, Switzerland: CERN; 1997

[46] Ghosh AK, Sampson WB, Bauer P, et al. Minimum quench energy measurements on single strands for LHC main magnets. IEEE Transactions on Applied Superconductivity. 1999;**9**(2): 252-256

[47] Calvi M, Busbar ITER. Design Studies, Swiss Federal Institute of Technology. V1.3, Lausanne: Center for Research in Plasma Physics; 2010

[48] Bauer P, Bessette D, Calvi M, Nannini M. ITER Busbar Thermohydraulic Anal-Ysis, ITER IDM 4D5SM9

[49] Zanino R, Bottura L, Marinucci C. A comparison between 1-and 2-fluid simulations of the QUELL conductor. IEEE Transactions on Applied Superconductivity. 1997;**7**(2):493-496

Chapter 7

Main Challenges of Heating Plasma with Waves at the Ion Cyclotron Resonance Frequency (ICRF)

Guillaume Urbanczyk

Abstract

Of all the techniques used for heating plasmas in fusion devices, waves in the Ion Cyclotron Resonance Frequency (ICRF $\sim$ MHz) continue to be exceptionally advantageous and unique insofar as it enables to deposit of power directly on ions in the core, significantly enhancing fast ion population together with fusion reaction products. However, because of the multiple inherent challenges—such as matching robustness, antenna design, wave coupling efficiency, wave propagation, wave absorption, and plasma surface interactions due to radiofrequency (RF) sheath excitation—ICRF is often one of the most complex heating systems to implement successfully. This chapter provides a brief introduction of these challenges and their respective underlying physics, together with examples of both simulations and experimental results from various tokamaks around the world. Finally, ICRF advantages and applications on present and future devices and perspectives of technological solutions are discussed and summarized.

Keywords: ICRF, wave coupling, RF sheath, plasma surface interactions

1. Introduction

The ICRF electrostatic wave is first produced by a generator called a tetrode, which is essentially a multistage amplifier of the power composed of a cascade of electron tubes. Each stage of which increases the power by approximately an order of magnitude, from watts to megawatts [1].

High-power waves enter the coaxial line of characteristic impedance Z_0, that matches the output impedance of the generator; typically, 50Ω or 30Ω as in the case of the Full-Tungsten Environment Superconducting Tokamak (WEST) [2] which **Figure 1** shows an overview of the ICRF system with all key elements.

1. Electrostatic waves then propagate towards the antenna and cross at some point the so-called *matching* unit which ensures a smooth transition between the impedance of the coaxial line Z_0 and the one of the antenna straps Z_S to avoid

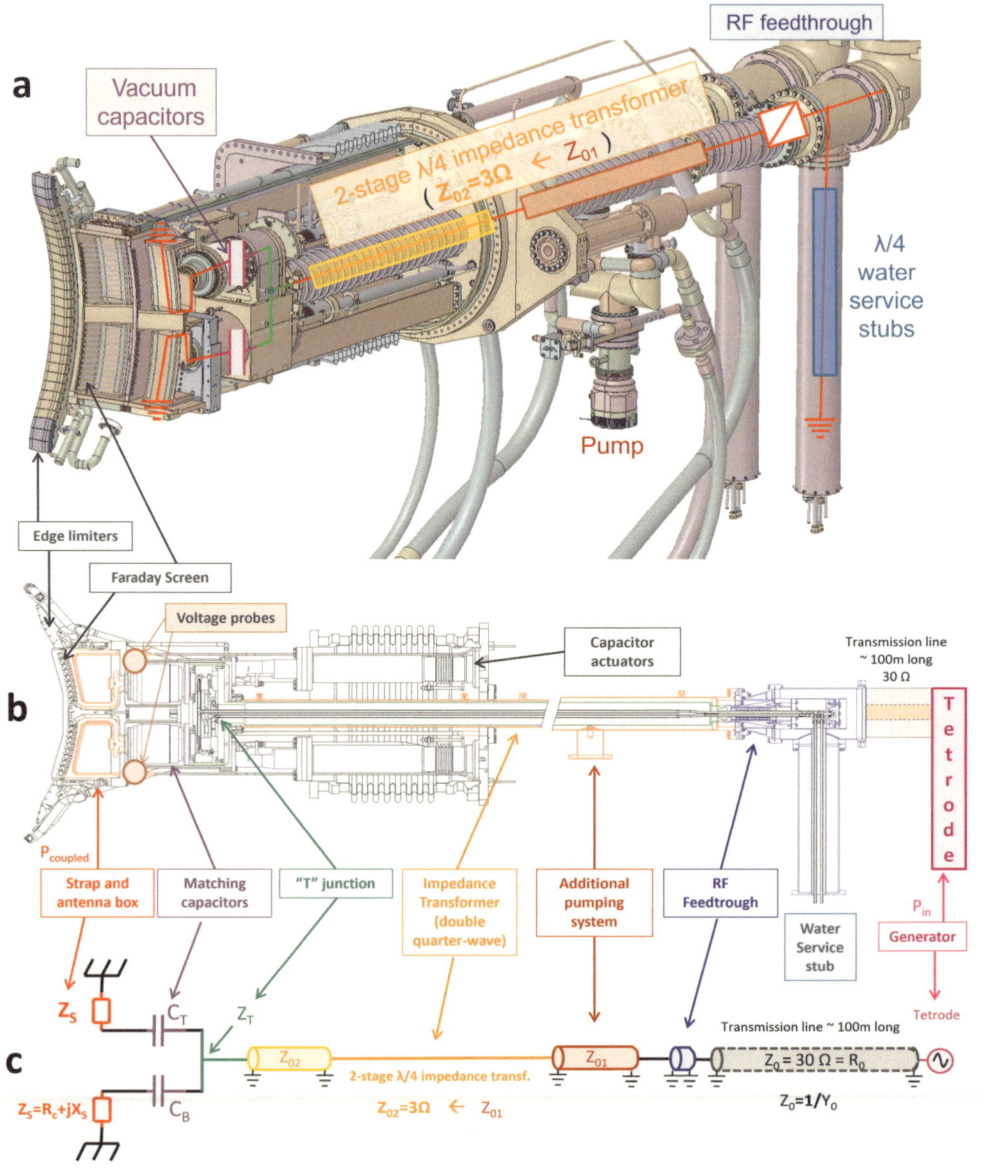

Figure 1.
ICRF system seen in 3D (a) and from profile (b), with a schematic of the key components of the RF circuit (c) from the generator to the antenna (case of WEST tokamak) adapted from [3].

power reflection. If successfully matched, the wave induces a current on the strap at the extremity of the transmission line.

2. This current oscillating at the ion cyclotron frequency (one the order of MHz), results in the excitation of electromagnetic ICRF waves and their *coupling* to the low-density edge plasma.

3. These *waves* must then propagate from the edge up to the resonance layer in the core, where they can be *absorbed by the ions*. Energy is then spread to all charged particles through collisions, resulting in *plasma heating*.

These three aspects, namely the *matching*, wave *coupling*, and *absorption*, are the key points mandatory for the successful operation of an ICRF system. Section 2 provides the physics background for each of these steps, while section 3 explains what happens in practice due to all of the imperfect conditions and introduces the challenges inherent in ICRF operation; power reflection, arcing, Radiofrequency (RF) sheath excitation, erosion, impurity generation, and plasma contamination. Despite all of these challenges, the fourth and final sections present the key assets that still make ICRF an attractive auxiliary heating system and some perspectives on solutions worthy of further investigation.

2. The key 3 steps of ICRH (Ion Cyclotron Resonance Heating)

2.1 RF circuit matching

Matching the RF circuit consists of building smooth transitions between the purely real impedance of the transmission line ($Z_0 = R_0 \approx 10 \rightarrow 50\Omega$) and the complex impedance of the antenna straps ($Z_S = R_C + jX_S$) which depends on the plasma. When the system is perfectly matched, the electrostatic wave flows from the generator to the straps without seeing any obstacle or being subject to any reflection. This can be achieved in many ways, relying on various components each with a variable impedance such as:

- *Single stub* also called *"trombones"* [4] in the Joint European Torus (JET) ITER-Like Antenna [5, 6] (ITER=International Tokamak Experimental Reactor), Alcator C-Mod [7] and the National Spherical Torus Experiment (NSTX) [8]
- *Double stubs loaded by shorted tunable capacitors* on TEXTOR [9] and analyzed for ASDEX-Upgrade [10]
- *Liquid stubs* arranged in a triple stub configuration in the Large Helical Device (LHD) [11] and the Experimental Advanced Superconducting Tokamak (EAST) [12]. The reactance of the short-circuited stubs of fixed geometrical length is varied by changing the height of the dielectric fluid (oil) inside the stubs
- *Fast Ferrite Tuners*: the reactance of a matching stub of fixed geometrical length is here modified by changing the magnetic characteristics of the ferrites which are disposed inside the stubs which were used in Alcator C-Mod [7] and ASDEX Upgrade [13]
- *Sliding Impedances* [14]
- *Single stubs in the Conjugate-T configuration* [15]
- *Tunable line-stretchers* (or *phase shifters*) in JET [6]
- *Tunable vacuum capacitors* in JET ITER-like antenna [16] and WEST ICRF antennas [17]. These are complex components ($Z_C = X_L - jX_C$), which self-inductance (X_L), and capacitance (X_C) can be tuned to match any load at the strap (cf. purple elements in **Figure 1**)

If the antenna is facing vacuum, the load is stable as it basically only corresponds to ohmic losses in the antenna. In this case, almost perfect matching can be achieved with the resistive part of the load remaining small ($Rc \approx 0.1\Omega$). In front of a plasma though, the load is larger than vacuum, so that Rc can increase by up to two orders of magnitude, in correlation with the efficiency by which the power is coupled.

In this case, however, the load (plasma) can change rapidly. This means Z_S is seldom if ever perfectly matched to Z_0, and that the *power* launched by the generator (P_{in}) is hardly exactly equal to the one *coupled by the antenna to the plasma*:

$$P_{coupled} = P_{in}\left(1 - |\Gamma|^2\right) \tag{1}$$

where Γ if the reflection coefficient defined as $\Gamma = (Z_s - Z_0) / (Z_s + Z_0)$.

The fraction of uncoupled power is therefore reflected in the transmission line, which can damage the generator if it is too large. This is avoided by introducing a quarter wavelength phase-shift in the matching circuit—for instance by tuning the height of the oil in EAST stub tuner or by using an impedance transformer in WEST—so that in the section between the generator and the matching circuit, only forward waves exist.

In the section between the matching circuit and the antenna though, coexist waves travelling both forward and backward, which superimpose on one another and give rise to a standing wave pattern. The maximum and minimum voltages of such waves are defined as follows:

$$V_{max} = (1 + |\Gamma|)\sqrt{2Z_0P_{in}} = \sqrt{2Z_0P_{coupled}VSWR} \tag{2}$$

$$V_{min} = (1 - |\Gamma|)\sqrt{2Z_0P_{in}} = \sqrt{2Z_0P_{coupled}/VSWR} \tag{3}$$

And the ratio of these voltages is the so-called *Voltage Standing Wave Ratio*:

$$VSWR = \frac{1 + |\Gamma|}{1 - |\Gamma|} \tag{4}$$

It follows that VSWR grows from 1 in the ideal perfectly matched "flat-line" case ($V_{max} = V_{min}$), to infinity when almost all the power is reflected. If the VSWR is too large, the difference of electrical potentials in the lines becomes so high that there is a significant risk of arcing. To prevent such deleterious events, operational safety limits are set on the voltage (V_{max}) and the current (I_{max}), above which the power shuts down.

$$I_{max} = (1 + |\Gamma|)\sqrt{2Y_0P_{in}} = \sqrt{2Y_0P_{coupled}VSWR} \tag{5}$$

with $Y_0=1/Z_0$. We now understand how crucial it is to transport the power along well-matched lines. It naturally follows that minimizing length of the lines with these undesirable effects by placing the matching system as close as possible to the antenna, will be a key aspect of the system's efficiency (cf. section 2.5 of [18]).

In addition, *decouplers* are used to prevent the mutual coupling of adjacent lines which influence each other due to the proximity of the straps to one another. This effect can make a source behave as a receiver and endanger its generator. Decouplers are used in DIII-D [19], EAST [20], Alcator C-Mod [7], NSTX [8] and will be used on ITER [21]. This is also important to equalize the voltages at the inputs of the array and guarantee a homogeneous excitation over the whole surface of the antenna's front face, which can otherwise have deleterious effects on local fields and RF sheath excitation. Strap-decoupling can also be improved by separating them with a septum in the antenna box to reduce mutual influence.

Excellent explanations of the topic can further be found in the fourth chapter of [18] with more detailed calculations and specific applications to the case of WEST. Once well matched, the system and the antenna are then ready to launch waves which coupling to the plasma remains to be optimized.

2.2 Wave coupling

The front of the antenna faces the plasma inside the vessel. As the power emitted from the generator reaches the antenna, it induces an oscillating current I_{RF} along the strap as highlighted in red in **Figures 1b** and **2**, which excites ICRF waves at the edge of the plasma, as efficiently as the load or the coupling resistance (R_c) is high.

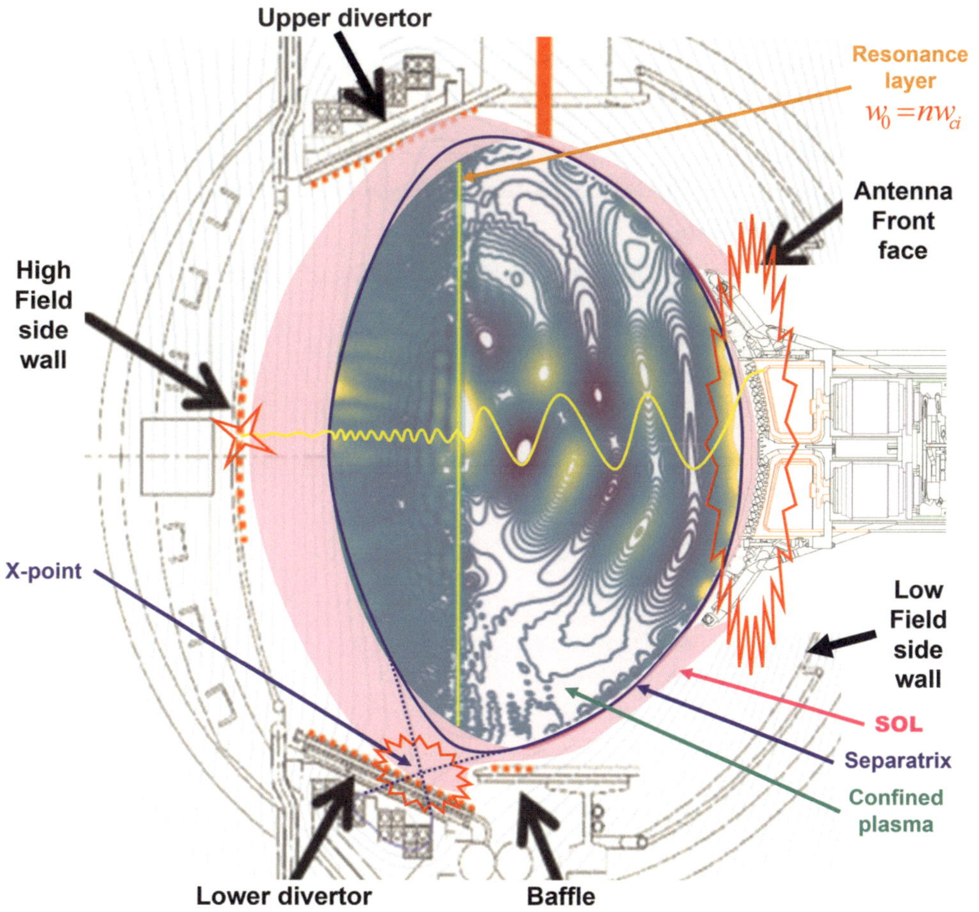

Figure 2.
Poloidal cut of a tokamak vacuum vessel, with ICRF waves excited by the antenna, propagating in the well-confined region of the plasma, partially absorbed and mode converted near the resonance layer.

$$P_{coupled} = \frac{1}{2} R_c I_{RF}^2 \quad (6)$$

The problem in coupling ICRF waves at the edge, lies in the fact that to avoid exposure to dense and hot plasma—which would induce unacceptably large heat loads—antennas are retracted away and sit in the Scrape-Off Layer (SOL), a low-density plasma (pink regions in **Figures 2** and **3**) where the Fast Wave (FW) is generally not propagative but evanescent ($k_\perp^2 < 0$). A wave is evanescent when the density is smaller than its cutoff density (n_{co}). Therefore, the larger the distance from the strap to the cutoff layer ($D_{Strap\text{-}co}$), the lower the coupling efficiency and vice versa. To give a rough idea, 1 cm increase of $D_{Strap\text{-}co}$ generally results in about 20% drop of the coupling efficiency, meaning that about 1/5 of the power vanishes every centimeter until reaching densities above n_{co}. In addition, the nature of the spectrum excited by the antenna, according to its geometry (straps width, height and spacing), and in particular, the value of the parallel wave vector ($k_{//}$) corresponding to the main components of the spectrum, plays a key role in the coupling process:

$$R_c = \frac{Z_0}{VSWR} \sim R_0 \exp\left(-2 < k_{//} > D_{Strap-co}\right) \quad (7)$$

One can observe, in **Figure 3**, how this parameter typically influences the fast-wave cutoff density which drops by over an order of magnitude as $k_{//}$ decreases from 18 down to 6 m^{-1}.

ICRF waves comport two modes called *fast* and *slow* waves in reference to their respective group velocity (details of mathematical calculations can be found in the second chapter of [22]):

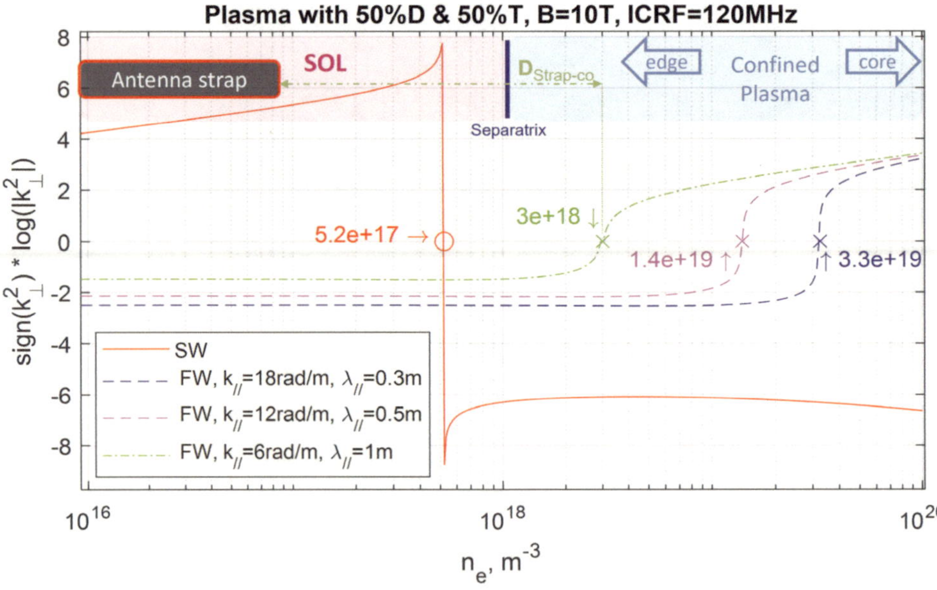

Figure 3.
Properties of 120MHz ICRF fast (FW—dashed lines) and slow (SW—solid line) wave modes in a realistic fusion plasma with similar proportions of deuterium and tritium confined by a magnetic field of 10 tesla. Waves are propagative when the sign of $k_\perp^2$ *is positive, and evanescent elsewhere. Cutoff densities of each branch are written explicitly on the graph.*

- *Fast Waves* (FW) are evanescent at low densities typical of the plasma edge and propagative for densities characteristic of the plasma core (**Figure 3**). FW have wavelengths on the order of few meters and an electric field perpendicular to the magnetic field. Therefore, when talking about coupling maximization, it is common to implicitly refer to this mode, because it is the one suited to heating the ions which rotate perpendicularly around the magnetic field lines.

- *Slow Waves* (SW) are propagative at very low densities characteristic of the outer most "far" SOL ($n_e < 5.2\ 10^{17} m^{-3}$ in **Figure 3**) and evanescent above. SW have millimetric wavelengths and an electric field parallel to the magnetic field. Because the current straps are not perfectly perpendicular to the tilted magnetic field lines, SW are parasitically excited and are often responsible for exciting the RF-sheath, leading to deleterious plasma surface interactions. The role the faraday screen (**Figure 1a**) with bars in principle parallel to the magnetic field, is intended to locally increase the conductivity in the parallel direction to screen the SW in front of the straps. Unfortunately, we will see in the next section that this is not enough.

At this point, one should remember when hearing people say that they are trying to "optimize the coupling of an ICRF antenna", that they are basically trying to maximize its coupling resistance (R_c), either by reducing $D_{Strap\text{-}co}$ or $k_{//}$ (Eq. (7)).

$\mathbf{D_{Strap\text{-}co}}$ can be minimized in different ways [23]:

- Put the *antenna closer to the plasma*, but this will come with stronger heat loads and usually impurity production

- *Globally increase the density* of the plasma, but this is limited and associated with an increase of the power lost in radiations

- *Locally increase the density* by fueling the plasma from valves magnetically connected to the antenna and ideally as close as possible to the middle plane [24, 25]

- *Increase the frequency* of the waves, which must remain compatible with an efficient scenario for heating the plasma (discussed in the next section)

$\mathbf{k_{//}}$ can be minimized at two different stages:

- While designing the antenna, basically by increasing the space between adjacent straps, but this often limited by the available space in the torus

- During the experiments, by reducing the phase difference of the currents on the adjacent straps (ϕ). Most of the time however, this trick is first limited by the currents induced on the surrounding passive structures, which play an important role in plasma surface interactions [26]. These currents basically tend to add up when currents are in phase (monopole ϕ=0°) instead of compensating each other for largest phasing (dipole ϕ=180°) [27]. Secondly, the larger $k_{//}$ the more efficient the wave absorption [28].

Once efficiently coupled to the plasma, wave absorption remains to be optimized.

2.3 Wave absorption

Waves that propagate towards the center of the plasma finally reach the resonance layer where ions frequency (w_0) corresponds to a multiple of the wave frequency (nw_{ci}), enlightened by the vertical yellow line in **Figure 2**. In this region, the frequency of the wave coincides with the fundamental or harmonics of the cyclotron frequency of ion species to be heated. One can tune the wave frequency, the magnetic field profile, and the plasma composition to pick the desired heating scheme among a variety of techniques [29] by matching the Doppler-shifted wave frequency with the ion cyclotron harmonics:

$$\omega = n\omega_{ci} + k_{//}v_{//,I} \tag{8}$$

- *Heating a minority at its fundamental resonance* (n=0): such as hydrogen in a deuterium plasma (H-D) with a small concentration of hydrogen, typically $n_H/n_H+n_D<15\%$. So far, minority heating is the most routinely used in most devices. However, while it has long been applied on two-ion species plasmas, it has been found that it can also be applied with great efficiency in plasmas with three (or more) ion species.

- The *three-ions scheme* also consists in heating the fundamental of the minority specie such as a little helium in a deuterium tritium fusion plasma (ex: (He^3)-DT), but here the resonance location of the minority must fall in between the resonances of both others majority species [30], where the amplitude of the left-handed polarization (E_+) can reach very high values and lead to strong diffusion (D) of the wave power:

$$D\propto|J_0|^2|E_+|^2 \tag{9}$$

- *Heating a majority at its harmonic resonance* (n>0): this can be an alternative to the minority heating typically if it proves challenging to accurately controlling the isotopic ratio. However, this scenario becomes truly effective only when a population of fast ions is created to improve power absorption. This process usually relies on harmonic heating. While generating a population of fast ions large enough to boost up the absorption, it can also be directly injected by a Neutral Beam Injector (NBI), hence the so-called synergy between NBI and ICRH (cf. experimental [31] and simulation [32] results in JET tokamak). Unfortunately, not only NBI is not available in all devices, but if the confinement is not good enough, for instance in case of low plasma current, this method is counteracted by fast-ions losses.

3. Challenges behind each step

3.1 Real-time matching

In practice, before main experiments start, ICRF operators usually "prepare the matching" and pre-set the parameters (ex: Oil level in the stub tuner in EAST, capacitance of the capacitors in WEST ...) with the antenna facing vacuum. Later, the matching is adjusted with the antenna facing the plasma, one of the many reasons why campaigns always include a period for commissioning systems at the beginning. The system starts with low power in case of sources of mismatch or poor coupling efficiency to avoid the risk of arcing due to large fractions of power reflected, and gradually raise it up to power levels relevant for the experiments (MW order). One of the key goals during the experiments is to further improve plasma confinement, often leading to operate in H-mode [33]. H-mode plasmas represent a great challenge for all RF circuit matching, first because during the L-H transition, a pedestal forms at the edge where the density typically drops with sharper gradient compared to L-mode plasmas, lowering the coupling efficiency (R_c) sometimes by half. But also, the so-called Edge-Localized Modes (ELMs) induce important load variations through transient bursts, basically as large as their frequency is low (from kHz range for the smallest type-III ELMs down to tens of Hz for the largest type-I ELMs with R_c suddenly increasing by factors up to 5 within less than a millisecond. One can further show for a strap matched by a 2-port matching unit, if the matching is not reconfigured, the VSWR will raise in similar proportions as R_c (cf. calculation details in section 2.7 of [18]), risking provoking arcs either in the transmission lines [34] or in the antenna box [35]. All this motivates real-time control of the impedance matching that can be achieved with:

- *Capacitors*, like in Tore Supra Classical antennas [36] and JET-ILA [37]. Amplitudes of the incident and reflected voltages and their phase shift are typically measured either on the transmission lines or ideally as close as possible to the antenna, and mixed to provide error signals driving the capacitors
- *Conjugate-T* and similar principle as with capacitors [13]
- *Fast Ferrite Tuners*, like in Alcator C-mod antennas [7]
- Impedance matching by real-time controlling the *generator's frequency* [4]
- *Decouplers* and *double stub* in ITER [21]

Various algorithms can then be used to automatically tune parameters of interest for each component. However so far, no controller allows tackling sub millisecond variations typically induced by ELMs, hence the need for so-called *load resilient* matching schemes [38].

3.2 Load resilience

Without load resilience, an ICRF system would stop delivering power every time its VSWR would exceed some safety value, which would basically make it incompatible with steady operation during a discharge with ELMs. A very convincing comparison between resilient and non-resilient system can be found in Figure 13 of [13].

Among existing load-resilient schemes, one can quote:

- 3dB *hybrid coupler* diverting the reflected power to a dummy load, such as the lossy network in ASDEX-Upgrade [39]
- *Conjugate-T* concept, connecting pairs of straps, with various designs:

- *External* conjugate-T relying on line-stretchers in JET [13] and EAST [40]
- *Internal* conjugate-T relying on capacitors in JET-ILA [5], Tore Supra [41] and WEST [17]
- *Stub* conjugate-T in Alcator C-Mod [15]

More details can be found on the conjugate-T under section 4 of the fourth chapter of [18] with detailed mathematical calculations and applications to the internal conjugate-T of WEST and experimental results in [42].

- *Travelling wave antenna* concept [43] which in case of load drop, takes profit from the mutual coupling between straps to let the wave propagate along the array and recirculate, making it inherently load resilient. This concept has been used in DIIID [44, 45] for driving current and envisaged for heating plasmas in WEST [46]. While never tested for heating tokamak plasmas, the concept somehow lost its appeal, essentially due to the integration challenges it raises, that become even more problematic in the perspective of fusion reactors which must maximize the surface of the tritium breading blanket and therefore focus on heating systems with large power densities.

Note that in most of the cases, load resilient solutions are combined with matching units, so that the problem is most often tackled from the generator perspective, with first goal to deliver constant power despite load fluctuations at the antenna.

After having explained the importance of operating with a well-matched system, we will now place ourselves from the inner vessel perspective and further discuss the challenges that concern an ICRF antenna facing a fluctuating plasma. There the main goal will rather be to prevent the coupling resistance from dropping under some value, not only critical for the VSWR and arcing in the transmission lines, but also for the excitation of near fields and RF sheath.

3.3 RF-sheath and potential rectification

The formation of a sheath on any component facing the plasma is a natural phenomenon [47]. The sheath is a thin layer (usually few millimeters wide) which forms on the surface of all materials in contact with the plasma. Across this layer, a separation of the charges (therefore an electric field) makes it possible to preserve the ambipolarity of the fluxes of charged particles from the plasma towards a surface (cf. **Figure 4**). Since the mass of electrons is much smaller than that of ions, their speed is much greater. Therefore, when exposed to a plasma, the surface of materials becomes negatively charged, repelling electrons, and attracting ions approaching below a *Debye length* ($\lambda_{De} = \sqrt{\varepsilon_0 k_B T_e / n_e e^2}$) with ε_0 and k_B respectively electric and Boltzmann constants, T_e and n_e respectively the electron temperature and density and the charge of the electron $e = 1.6 \times 10^{-19}$ coulombs.

The properties of the sheath can undergo substantial changes depending on several parameters that can typically influence its electric field and width δ_{sheath}:

$$\delta_{sheath} = \lambda_{De} \left(\frac{e.V_{DC}}{k_B T_e} \right)^m \tag{10}$$

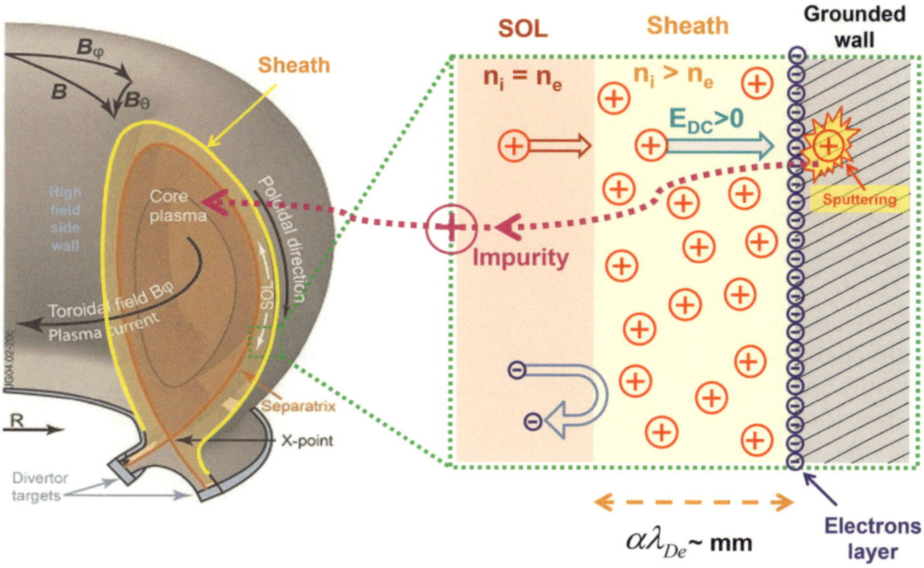

Figure 4.
Sketch of a sheath layer forming on a material facing the plasma.

with V_{DC} the DC potential and m a parameter that typically changes between 2/3 and ¾ depending on the incidence angle of magnetic field lines on materials and the collisional properties of the sheath as explored through Particle in Cell simulations in [48]. In general, the electric field accelerates the ions towards the wall, leading to an increase of the sputtering and the production of impurities (**Figure 4**) in proportions usually comparable to the increase in electron temperature:

$$V_{\text{Thermal sheath}} = 3k_B T_e/2 \tag{11}$$

On the other hand, in the presence of ICRF heating, this phenomenon is further aggravated by the so-called *rectification* of the oscillating potentials associated with the *slow wave* carrying electric field in the direction parallel to the magnetic field lines (cf. red curve in **Figure 3**). Let us now express the *current instantly conducted to any object exposed to the plasma* as:

$$\tilde{I}(t) = I_{sat}^{+}\left(1 - \exp\left(\frac{e}{k_B T_e}\left(\phi_f - \tilde{V}(t)\right)\right)\right) \tag{12}$$

with I_{sat}^{+} the ion saturated current, ϕ_f the floating potential and $\tilde{V}(t) = V_{DC} + V_{RF}\cos(\omega_0.t)$ the sheath potential in presence of an RF wave. Then the RF ➔ DC rectification, specific to waves at the ion cyclotron frequency, results from the fact that the electrons (light compared to the ions) react instantaneously to the oscillating electric field of the wave (V_{RF}), while the ions react only to the *average value*.

$$\langle\tilde{I}(t)\rangle = \bar{I} = I_{sat}^{+}\left[1 - \exp\left(\frac{e}{k_B T_e}\left(\phi_f - V_{DC}\right)\right)I_0\left(\frac{eV_{RF}}{k_B T_e}\right)\right] \tag{13}$$

with I_0 a Bessel function. The consequence at the level of the components facing the plasma on the scale of an RF period, is the appearance of a DC current due to the privileged drift of the electrons. Thus, to compensate this current, the DC potential of the sheath must adjust and get biased by an extra potential V_b:

$$V_b = \frac{k_B T_e}{e} I_0 \left(\frac{e V_{RF}}{k_B T_e} \right) \tag{14}$$

The RF → DC rectification of the sheath therefore has the effect of increasing the potential drop (electric field) through the skin between the plasma and the wall, i.e., the acceleration of the ions towards the materials. This acceleration often boosts ions incident energies up to critical levels for exposed components (i.e., above sputtering yields threshold values). For deeper understanding of the topic, the reader is highly encouraged to refer to excellent tutorial [49] and reviews of key experimental [26–50] and modelling [51] results.

3.4 Physical sputtering yield

The other key process by which ICRF operation often enhances impurity production is the physical sputtering yield Y_{Eff}. This parameter represents the probability for an ion to sputter an atom from a target, given its charge, its incident energy, and their respective mass:

$$\text{Impurity outflux} = \Gamma_{\text{target atoms out}} = \Gamma_{\text{ion in}} Y_{\text{Eff}} \tag{15}$$

For example, the graph in **Figure 5** shows how the sputtering of a graphite (~carbon) target evolves for different bombarding species. The first important aspect of the sputtering yield is that it varies non-linearly with the energy of the incident ion. The second one is that only above a threshold energy of about 30eV, do the incident species start to sputter carbon atoms. The red curve also represents the self-sputtering of lithium which is sometimes used for material coating to prevent from important plasma contamination by high-Z impurities. One can see that a lithium coating will typically get eroded at much lower energy, while plasma facing components in graphite will require larger energies. We have here taken the case of carbon components, sometimes coated with lithium, exposed to various species, which is the case in EAST tokamak. Carbon however won't be used in future devices plasma facing components due to tritium retention. Tungsten has now become the most suitable candidate. In WEST for instance, antenna limiters are coated with tungsten, which sputtering follows similar trends (cf. **Figure 3** of [23]).

3.5 Poor coupling efficiency and near-field effects

RF-sheath excitation is the phenomenon proper to ICRF which is responsible for its peculiar complexity, due to the non-linear trade-off relation between the maximization of the coupling and the minimization of the interactions with the plasma. For any object in principle, moving away from the plasma results in a strict decrease of their interactions. However, for a classic ICRF antenna, moving away first induces a drop in its coupling efficiency, therefore an increase of excited fields (consistent with an increase of the VSWR in the transmission lines), which in turn can in critical cases result in a local but exponential increase of the RF sheath potential (cf. Figure 3 of [52]

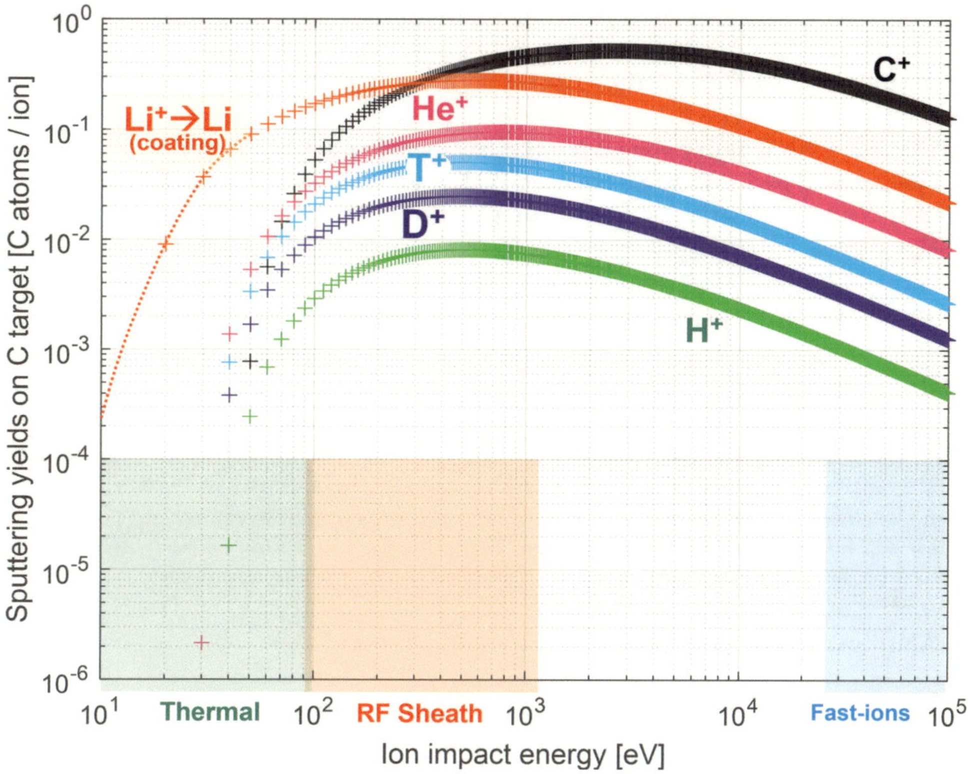

Figure 5.
Sputtering yield of various ion species on a target in graphite.

and section 3.3 of this chapter). In addition, the sputtering yield not only changes non-linearly, but also reaches its largest values for energies typically in the range of potential rectified by RF sheaths (cf. red region between 100eV and 1keV in **Figure 4**). It follows that in some cases, increasing the distance between an ICRF antenna and the plasma to reduce their interactions based on linear intuition, can unexpectedly cause an increase of impurity production from local sources, reacting to sharp increase of RF sheath potentials. These phenomena can be modeled in different ways as summarized in [53]. This behavior was also experimentally observed in WEST during a scan of the radial position of an ICRF antenna, where the floating potential measured by a reciprocating probe did not directly decrease as the antenna was retracted but first passed by a maximum for intermediate position (cf. Figure 4 in [50] or Figure 6.23 of [22] in open access). A wide variety of RF-sheath induced plasma surface interactions have also been observed in devices such as:

- EAST with the partial melting of metallic plates at the corners of an ICRF antenna, which precisely did not melt at the closest point to the plasma, but at the locations where simulations predict RF sheath excitation to be maximal (cf. Figure 14 in [54] or Figure 6.27 of [22] in open access)

- Tore Supra with large, enhanced potentials measured with retarded field analyzer [55] causing large heat loads on antennas [56]

- A LineAr Plasma Device (LAPD) using diagnostic magnetically connected to ICRF antenna along large number of reproducible discharges to provide great evidence of ICRF interaction process in the edge plasma [57]

- Alcator C-Mod with strong potentials measured with several different probes [58] caused by RF sheath excitation [59], leading to the increase of impurity sources and deleterious influence on plasma performance [60]

- ASDEX Upgrade with similar experiments [61] that have inspired a series of successful technological upgrades [62, 63] discussed hereafter in section 4

- JET is also subject to these effects [26, 64]

Interactions occurring at vicinity or in regions magnetically connected to the active antenna frame (red start with most branches around the antenna in **Figure 2**) are often classified as *near field effects.* Hence, interactions respectively taking place elsewhere will be conveniently referred to as *far field effects*[1].

3.6 Abnormal wave absorption and far-field effects

Following the same logic, we will categorize interactions occurring in regions without magnetic connection to the active antenna as *far field effects*. These can be observed in unexpected locations when abnormal propagation and absorption of the wave take place. For instance, if the absorption efficiency is low, the power launched in the plasma is not fully absorbed at the first pass, and a non-negligible fraction of unabsorbed power propagates and reflects on farther objects such as the divertor or the inner wall as represented in **Figure 2**. In these cases, not only heating performance drop, but RF-sheath can also potentially become excited in global fashion. Such effects were well-observed in

- EAST *high field side wall* (cf. red star with 4 branches in **Figure 2**) when the antenna operated in monopole phasing, causing a drop in the absorption efficiency and an increase of fields excitation at the inner wall [54]

- WEST *divertor* if too low hydrogen ratio leads to poor wave absorption [65]

- NSTX *divertor* where operating at high harmonics led to the excitation of surface waves [66] and substantial fraction of their power to be coupled to the edge filamentary structures parallel to the magnetic field lines [67], resulting in potential rectification at their extremity and deleterious interactions [68].

---

[1] Here, *near* and *far field effects* should not be confused with the near and far electromagnetic fields diffusion patterns around an object, but simply refer to the location where the effect occurs with respect to the ICRF antenna.

4. The beauty of ICRH: assets and technological perspectives

4.1 Assets

Despite the many challenges, ICRF has remained a very important tool for heating plasmas, with constant progress over the past fifty years [69]. ICRF is the only heating system that can directly deposit power onto the ions, allowing to generate fast ions and significantly boost fusion performance. While minority heating has long remained the most widely used method, this past two decades have been marked by the emergence of several modelling tools allowing to predict wave absorption efficiency [70] and power thermalization on the various species [71, 72]. These tools have allowed to explore much wider varieties of heating schemes [29]. Most promising ones were then tried experimentally, and their efficiency confirmed. One can typically quote scenarios based on the synergy with NBI [31, 32, 73] and these relying on three-ion scheme [74]. In particular, the three-ion scheme has gained increasing interest not only due to the efficiency in generating fast ions, but also for the operation flexibility it offers. Indeed, instead of relying on specific fraction of a minority specie—which can be difficult to maintain over operation—it proves much easier to keep control of rough proportions of two (or more) majority species and deposit power on a third very minor one which fraction is easier to control. Most promising scenario for fusion reactor D-T plasmas would therefore consist in depositing power most likely on ^3He, which has been successfully achieved in JET and Alcator C-Mod [30] and foreseen for future devices like ITER [74] and SPARC [75].

Another application of ICRH is the control of impurity population in the core. It has for instance been observed in JET, that for ICRF power above 4MW, heating H minority can result in high-Z impurity screening out of the core [76]. Furthermore, the three-ion scheme even allows to directly pump out an impurity by depositing part of the power on it. This was observed with beryllium in JET [29], with Argon on TFR [77] and Alcator C-Mod [78]. The same trick could further allow to pump out tungsten impurity by targeting the second harmonic of W^{56+} in SPARC's high temperature plasmas [78], a powerful tool to prevent tungsten accumulation in the core.

4.2 Technological perspectives

A few research and development ideas for improving current design of antennas yet remain to be tested. One can quote:

- *Travelling wave antennas* [43] could improve the reliability of ICRF systems thanks to its inherent load resilience and much larger coupling capacities, offering the possibility to place the antenna 10cm further away from the plasma than a classic antenna and still be able to couple similar amount of power. This could have tremendous impact not only on the coupling efficiency but also on the interactions with the plasma and its contamination by metallic impurity. The disadvantage of this concept is its low power density due to the large number of straps (>5). This makes any design hardly compatible with fusion reactor worried of maximizing the surface of the tritium breeding blanket. But maybe some tricks have not yet been thought about ...

- *Active limiters* could allow compensating actively the image currents induced on limiters by nearby current straps. Relying on the concept of proximity effect [27], these currents can be cancelled by powering straps with appropriate phasing and power ratio, as was shown in ASDEX-Upgrade 3-strap antennas [63] and Alcator C-Mod 4-strap antennas [79]. However, operating the ICRF system with optimal settings is not only challenging in terms of control, but mostly limiting in terms of power density. *Active limiters* would then consist in applying a fraction of the power sent on the straps, directly on the limiters. This could allow reducing the local impurity sources, by canceling off the local currents that tend to be responsible for RF sheath excitation. In addition, such solution would offer two key advantages:

    - *more flexibility* during ICRF operation by opening the possibility to routinely tune the antenna *phasing* without aggravating interactions with the plasma. In other words, this would allow operating with different antenna spectra to cope with plasma changes regardless of near-field effects

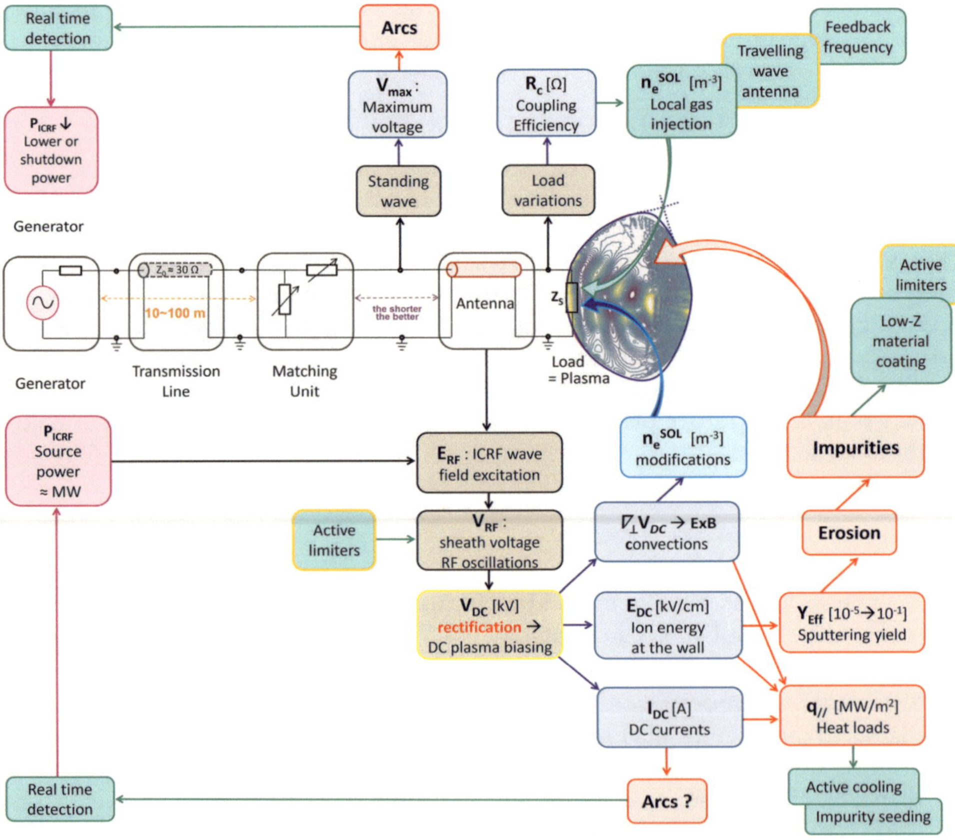

Figure 6.
Summary of the key components of an ICRF system, and a selection of the main phenomena that can occur in various location. Black boxes enlighten phenomena inherent to ICRF powering. Blue boxes show the physical parameters that are most directly influenced. Red boxes show the potentially deleterious consequences of parameters changes. Green boxes with green and orange borders show the main solutions respectively existing and to be explored.

- *maximize power density*. While this is already quite important in nowadays experimental tokamaks, it will be a critical aspect on future fusion reactors, which aim at maximizing the surface of the tritium breeding blanket for producing fuel and extracting the energy radiated by the plasma to generate electricity. Simultaneously satisfying requirements on plasma heating, fuel production and energy exhaust, pushes towards more compact designs with larger power density currently hardly compatible with low level of impurity production. Active limiters could therefore be a major technological solution.

5. Conclusion

Despite all challenges, heating fusion plasmas with waves in the ion cyclotron range of frequency is still very attractive. While several technological solutions remain to be explored, a variety of existing ones are already routinely used and allow successful operation in many devices around the globe. A summary of the key components of an ICRF system, and a selection of the main phenomena that can occur in various location is provided in **Figure 6**. To conclude on the perspective of fusion reactors, while ICRF may not be the optimal solution in terms of power density, note that as the intensity of the magnetic field due to progress made on low temperature superconductors and the size of the devices will increase, ICRF is the only heating system which difficulty to implement will not be affected.

For further reading of on this topic, the reader is highly encourage to refer to a review of the "Recent Progress in ICRF in Magnetic Confinement Fusion" [80] and references therein.

Acknowledgements

Laurent Colas, Julien Hillairet, Karl Vulliez, Annika Ekedahl, and more globally WEST and EAST ICRF Teams are warmly acknowledged for their efforts in making experiments possible and the numerous fruitful discusssssions. I also want to thank a lot Wouter Tierens, Mari Usoltceva, Vladimir Bobkov, and the rest of ASDEX ICRF team, Ernesto Lerche from ERM and colleagues from the 107 team of Institut Jean Lamour for all the things we collaborated over the past few years.

Author details

Guillaume Urbanczyk
Institute of Plasma Physics, Chinese Academy of Sciences, Hefei, China

*Address all correspondence to: guiguiurban@hotmail.com

References

[1] Faugel H, Bobkov V, Fünfgelder H. The prospects of ICRF generators at ASDEX upgrade. AIP Conference Proceedings. 2020;**2254**:070006. DOI: 10.1063/5.0013529

[2] Bourdelle C et al. WEST Physics Basis. Nuclear Fusion. 2015;**55**:063017. DOI: 10.1088/0029-5515/55/6/063017

[3] Vulliez K, Chen Z, Ferlay F, Winkler K, Helou W, Hillairet J. The mechanical structure of the WEST ion cyclotron resonant heating launchers. Fusion Engineering and Design. 2015; **96–97**:611-615. DOI: 10.1016/j.fusengdes.2015.06.050

[4] Wade TJ, Jacquinot J, Bosia G, Sibley A, Schmid M, et al. Development of the JET ICRH plant. Fusion Engineering and Design. 1994;**24**:23-46. DOI: 10.1016/0920-3796(94)90035-3

[5] Durodié F, Dumortier P, Blackman T, Wooldridge E, Lerche E, Helou W, et al. ITER-like antenna for JET first results of the advanced matching control algorithms. Fusion Engineering and Design. 2017;**123**(November):253-258. DOI: 10.1016/j.fusengdes.2017.05.113

[6] Monakhov I, Graham M, Blackman T, Dowson S, Durodie F, Jacquet P, et al. Design and operations of a load-tolerant external conjugate-T matching system for the A2 ICRH antennas at JET. Nuclear Fusion. 2013;**53**(8):083013. DOI: 10.1088/0029-5515/53/8/083013

[7] Lin Y et al. ICRF antenna matching systems with ferrite tuners for the Alcator C-Mod tokamak. Fusion Engineering and Design. 2015;**100**: 239-248. DOI: 10.1016/j.fusengdes.2015.06.022

[8] Swain DW et al. Loading and asymmetry measurements and modeling for the national spherical torus experiment ion cyclotron range of frequencies system. Fusion Science and Technology. 2003;**43**:503-513. DOI: 10.13182/FST03-A297

[9] Durodié F, Vervier M. First results of the automatic matching device for textor's ICRH system, Fusion Technology. 1992. pp. 477-480. DOI: 10.1016/B978-0-444-89995-8.50088-6. ISBN 9780444899958

[10] Hangan D-M et al. Analysis of dynamic matching networks for the ICRF system at ASDEX upgrade. Fusion Engineering and Design. 2011;**86**: 736-741. DOI: 10.1016/j.fusengdes.2011.02.001

[11] Saito K et al. Real-time impedance matching system for ICRF heating in LHD. Fusion Engineering and Design. 2008;**83**:245-248. DOI: 10.1016/j.fusengdes.2008.01.017

[12] Zhao Y-P et al. EAST ion cyclotron resonance heating system for long pulse operation. Fusion Engineering and Design. 2014;**89**:2642-2646. DOI: 10.1016/j.fusengdes.2014.06.017

[13] Braun F et al. ICRF system enhancements at ASDEX upgrade. Fusion Engineering and Design. 2001; **56–57**:551-555. DOI: 10.1016/S0920-3796(01)00274-5

[14] Lamalle PU et al. Commissioning of the wideband matching system for ICRH of ELMy JET Plasmas. AIP Conference Proceedings. 2001;**595**:118

[15] Parisot A. Design of an ICRF fast matching system for the Alcator C-Mod tokamak. PhD Thesis. 2004. Available from: https://dspace.mit.edu/bitstream/handle/1721.1/93763/

04RR002_full.pdf?sequence=1&isAllowed=y

[16] Durodié F et al. Physics and engineering results obtained with the ion cyclotron range of frequencies ITER-like antenna on JET. Plasma Physics and Controlled Fusion. 2012;**54**:074012. DOI: 10.1088/0741-3335/54/7/074012

[17] Helou W, Colas L, Hillairet J, Milanesio D, Mollard P, Argouarch A, et al. Radio-frequency electrical design of the WEST long pulse and load-resilient ICRH launchers. Fusion Engineering and Design. 2015;**96–97** (October):473-476. DOI: 10.1016/j.fusengdes.2015.01.005

[18] Helou W. Design and operation of antennas at the ion cyclotron and lower hybrid range of frequencies for nuclear fusion reactors. PhD Thesis. 2017. Available from: http://www.theses.fr/2017AIXM0395

[19] Goulding RH, Hoffman DJ, Ryan PM. Power compensators for phased operation of antenna arrays on JET and DIII-D. AIP Conference Proceedings. 1994;**289**:351. DOI: 10.1063/1.44958

[20] Gen C, Mao YZ, Zhao YP, Yuan S, Zhang XJ, Qing CM. Development and test of decoupler for ICRF antenna in EAST. Fusion Engineering and Design. 2016;**107**:32-37. DOI: 10.1016/j.fusengdes.2016.04.010

[21] Grine D et al. Summary and results of the study of the hybrid matching option implementation of the ITER ICRH system. Fusion Engineering and Design. 2012;**87**:167-178. DOI: 10.1016/j.fusengdes.2011.12.006

[22] Urbanczyk G. Interaction of High-Power waves with the plasma periphery of WEST & EAST tokamaks. PhD Thesis. 2019. Available from: https://hal.univ-lorraine.fr/tel-02517768/document

[23] Urbanczyk G, Colas L, Hillairet J, Lerche E, Fedorczak N, et al. RF wave coupling, plasma heating and characterization of induced plasma-material interactions in WEST L-mode discharges. Nuclear Fusion. 2021;**61**: 086027. DOI: 10.1088/1741-4326/ac0d11

[24] Urbanczyk G, Zhang XZ, Colas L, et al. Optimization of discharges with ion cyclotron range of frequencies using local gas injection in EAST. Nuclear Fusion. 2019;**59**:066023. DOI: 10.1088/1741-4326/ab1610

[25] Zhang W, Jacquet P, Lerche E, Bilato R, Bobkov V, et al. 3D simulations of gas puff effects on edge plasma and ICRF coupling in JET. Nuclear Fusion. 2017;**57**:056042. DOI: 10.1088/1741-4326/aa6817

[26] Bobkov V, Aguiam D, Bilato R, Brezinsek S, Colas L, et al. Impact of ICRF on the scrape-off layer and on plasma wall interactions: From present experiments to fusion reactor. Nuclear Materials and Energy. 2019;**18**:131-140. DOI: 10.1016/j.nme.2018.11.017

[27] Colas L, Lu L-F, Křivská A, et al. Spatial proximity effects on the excitation of Sheath RF voltages by evanescent slow waves in the ion cyclotron range of frequencies. Plasma Physics and Controlled Fusion. 2017;**59**: 025014. DOI: 10.1088/1361-6587/59/2/025014

[28] Lerche E, Bobkov V, Jacquet P, Mayoral M-L, Messiaen A, et al. Recent experiments on alternative dipole phasing with the JET A2 ICRF antennas. AIP Conference Proceedings. 2009;**1187**: 93. DOI: 10.1063/1.3273845

[29] Lerche E, Van Eester D, Jacquet P, Casson F, et al. ICRH options for JET-ILW DTE2 operation. AIP Conference Proceedings. 2020;**2254**:030007. DOI: 10.1063/5.0013530

[30] Kazakov Y, Ongena J, Wright J, et al. Efficient generation of energetic ions in multi-ion plasmas by radio-frequency heating. Nature Physics. 2017;**13**: 973-978. DOI: 10.1038/nphys4167

[31] Kirov KK, Kazakov Y, Nocente M, Ongena J, et al. Synergistic ICRH and NBI heating for fast ion generation and maximising fusion rate in mixed plasmas at JET. AIP Conference Proceedings. 2020;**2254**:030011. DOI: 10.1063/5.0014235

[32] Huynh P, Lerche EA, Van Eester D, Bilato R, et al. Modeling ICRH and ICRH-NBI synergy in high power JET scenarios using European transport simulator (ETS). AIP Conference Proceedings. 2020;**2254**:060003. DOI: 10.1063/5.0014240

[33] Keilhacker M et al. Confinement studies in L and H-type Asdex discharges. Plasma Physics and Controlled Fusion. 1984;**26**(1A):49-63. DOI: 10.1088/0741-3335/26/1A/305

[34] Bobkov V. Studies of high voltage breakdown phenomena on ICRF antennas. 2003. Available from: http://mediatum.ub.tum.de/doc/602957/document.pdf

[35] Bobkov V, Braun F, Hartmann DA, et al. Influence of ELMs on operation of ICRF antennas in ASDEX Upgrade. Journal of Nuclear Materials. 2005; **337–339**:776-780. DOI: 10.1016/j.jnucmat.2004.10.109

[36] Ladurelle L, Agarici G, Beaumont B, et al. First results of automatic matching system on Tore Supra ICRH antennas: fast matching network for ICRH systems. In: Proceedings of 19th SOFT. Lisbonne, Portugal; 1996. DOI: 10.1016/B978-0-444-82762-3.50118-X

[37] Durodié F, Dumortier P, Helou W, et al. Circuit model of the ITER-like antenna for JET and simulation of its control algorithms. AIP Conference Proceedings. 2015;**1689**(070013). DOI: 10.1063/1.4936520

[38] Noterdaeme J-M, Bobkov V, Brémond S, Parisot A, et al. Matching to ELMy plasmas in the ICRF domain. Fusion Engineering and Design. 2005;**74**: 191-198. DOI: 10.1016/j.fusengdes.2005.06.071

[39] Faugel H, Angene P, Becker W, Braun F, Bobkov V, et al. The ASDEX upgrade ICRF system: Operational experience and developments. Fusion Engineering and Design. 2005;**74**: 319-324. DOI: 10.1016/j.fusengdes.2005.06.268

[40] Yang H, Zhang XJ, Yuan S, Qin CM, et al. Experimental validation of the newly designed ICRF antenna on EAST. Private Communication

[41] Vulliez K, Argouarch A, Bosia G, et al. Validation of the load-resilient ion cyclotron resonance frequency antenna concept on Tore Supra plasmas. Nuclear Fusion. 2008;**48**:065007. DOI: 10.1088/0029-5515/48/6/065007

[42] Hillairet J, Mollard P, Colas L, Helou W, Urbanczyk G, et al. WEST actively cooled load resilient ion cyclotron resonance heating system results. Nuclear Fusion. 2021;**61**:096030. DOI: 10.1088/1741-4326/ac1759

[43] Ragona R, Messiaen A. Conceptual study of an ICRH traveling-wave antenna system for low-coupling conditions as expected in DEMO.

Nuclear Fusion. 2016;**56**:076009. DOI: 10.1088/0029-5515/56/7/076009

[44] Pinsker RI, Prater R, Moeller CP, deGrassie JS, et al. Experiments on helicons in DIII-D—investigation of the physics of a reactor-relevant non-inductive current drive technology. Nuclear Fusion. 2018;**58**:106007. DOI: 10.1088/1741-4326/aad1f8

[45] Van Compernolle B, Brookman MW, Moeller CP, Pinsker RI, Garofalo AM, O'Neill R, et al. The high-power helicon program at DIII-D: Gearing up for first experiments. Nuclear Fusion. 2021; **61**(11):116034. DOI: 10.1088/1741-4326/ac25c0

[46] Ragona R, Batal T, Hillairet J, Messiaen A, et al. Progress on the design of a DEMO high power ICRH travelling wave antenna mock-up to be tested on WEST. AIP Conference Proceedings. 2020;**2254**:070014. DOI: 10.1063/5.0014215

[47] Stangeby PC. The Plasma Boundary of Magnetic Fusion Devices (1st ed.). Boca Raton: CRC Press. 2000. DOI: 10.1201/9780367801489

[48] Moritz J, Heuraux S, Gravier E, Lesur M, Brochard F, et al. Sheath size and Child–Langmuir law in one dimensional bounded plasma system in the presence of an oblique magnetic field: PIC results. Physics of Plasmas. 2021;**28**:083501. DOI: 10.1063/5.0055790

[49] Myra JR. A tutorial on radio frequency sheath physics for magnetically confined fusion devices. Journal of Plasma Physics. 2021;**87**(5): 905870504. DOI: 10.1017/S002237 7821000878

[50] Colas L, Urbanczyk G, Goniche M, Hillairet J, Bernard J-M, Bourdelle C, et al. The geometry of the ICRF-induced wave–SOL interaction. A multi-machine experimental review in view of the ITER operation. Nuclear Fusion. 2022; **62**(1):016014. DOI: 10.1088/1741-4326/ac35f9

[51] Colas L, Lu LF, Jacquot J, Tierens W, et al. SOL RF physics modelling in Europe, in support of ICRF experiments. EPJ Web of Conferences. 2017;**157**: 01001. DOI: 10.1051/epjconf/201715701001

[52] Myra JR, D'Ippolito DA. Resonance cone interaction with a self-consistent radio-frequency sheath. Physical Review Letters. 2008;**101**:195004. DOI: 10.1103/PhysRevLett.101.195004

[53] Zhang W, Bilato R, Bobkov VV, et al. Recent progress in modeling ICRF-edge plasma interactions with application to ASDEX upgrade. Nuclear Fusion. 2021. DOI: 10.1088/1741-4326/ac38c8

[54] Urbanczyk G, Zhang XJ, Colas L, Dumont R, Tierens W, et al. Metallic impurity content behavior during ICRH-heated L-mode discharges in EAST. Nuclear Fusion. 2020;**60**:126003. DOI: 10.1088/1741-4326/abae82

[55] Martin K, Gunn JP, Heuraux S, Colas L, Faudot E, Jacquot J. Measurement of sheath potential in RF-biased flux tubes using a retarding field analyzer in Tore Supra tokamak. Journal of Nuclear Materials. 2013;**438**: S509-S512. DOI: 10.1016/j.jnucmat.2013.01.105

[56] Corre Y, Firdaouss M, Colas L, Argouarch A, Guilhem D, Gunn J, et al. Characterization of heat flux generated by ICRH heating with cantilevered bars and a slotted box Faraday screen. Nuclear Fusion. 2012;**52**:103010. DOI: 10.1088/0029-5515/52/10/103010

[57] Martin MJ, Gekelman W, Van Compernolle B, Pribyl P, Carter T. Experimental observation of convective cell formation due to a fast wave antenna in the large plasma device. Physical Review Letters. 2017;**119**:205002. DOI: 10.1103/PhysRevLett.119.205002

[58] Ochoukov R, Whyte DG, Brunner D, D'Ippolito DA, et al. ICRF-enhanced plasma potentials in the SOL of Alcator C-Mod. Plasma Physics and Controlled Fusion. 2014;**56**:015004. DOI: 10.1088/0741-3335/56/1/015004

[59] Wukitch SJ, LaBombard B, Lin Y, Lipschultz B, Marmar E, et al. ICRF specific impurity sources and plasma sheaths in Alcator C-Mod. Journal of Nuclear Materials. 2009;**390–391**: 951-954. DOI: 10.1016/j.jnucmat.2009.01.245

[60] Wukitch SJ, Lipschultz B, Marmar E, Lin Y, et al. RF plasma edge interactions and their impact on ICRF antenna performance in Alcator C-Mod. Journal of Nuclear Materials. 2007; **363–365**:491-497. DOI: 10.1016/j.jnucmat.2007.01.273

[61] Dux R, Bobkov V, Herrmann A, Janzer A, Kallenbach A, et al. Plasma-wall interaction and plasma behaviour in the non-boronised all tungsten ASDEX Upgrade. Journal of Nuclear Materials. 2009;**390–391**:858-863. DOI: 10.1016/j.jnucmat.2009.01.225

[62] Bobkov V, Balden M, Bilato R, Braun F, et al. ICRF operation with improved antennas in ASDEX Upgrade with W wall. Nuclear Fusion. 2013;**53**: 093018. DOI: 10.1088/0029-5515/53/9/093018

[63] Bobkov V, Aguiam D, Bilato R, Brezinsek S, Colas L, et al. Making ICRF power compatible with a high-Z wall in ASDEX upgrade. Plasma Physics and Controlled Fusion. 2017;**59**:014022. DOI: 10.1088/0741-3335/59/1/014022

[64] Colas L, Jacquet P, Bobkov V, Brix M, Meneses L, et al. 2D mappings of ICRF-induced SOL density modifications on JET. In: 45th EPS Conference on Plasma Physics Proceedings. http://ocs.ciemat.es/EPS2018PAP/pdf/O4.101.pdf

[65] Urbanczyk G, Colas L, Zhang XJ, Helou W, Hillairet J, et al. ICRH coupling optimization and impurity behavior in EAST and WEST. AIP Conference Proceedings. 2020;**2254**:030012. DOI: 10.1063/5.0018453

[66] Tierens W, Colas L, EUROfusion MST1 Team. Slab-geometry surface waves on steep gradients and the origin of related numerical issues in a variety of ICRF codes. Journal of Plasma Physics. 2021;**87**(4):905870405. DOI: 10.1017/S002237782100074X

[67] Tierens W, Zhang W, Manz P, et al. The importance of realistic plasma filament waveforms for the study of resonant wave-filament interactions in tokamak edge plasmas editors-pick Physics of Plasmas. 2020;**27**:052102. DOI: 10.1063/5.0007098

[68] Perkins RJ, Hosea JC, et al. High-harmonic fast-wave power flow along magnetic field lines in the scrape-off layer of NSTX. Physical Review Letters. 2012;**109**:045001. DOI: 10.1103/PhysRevLett.109.045001

[69] Noterdaeme JM. Fifty years of progress in ICRF, from first experiments on the model C stellarator to the design of an ICRF system for DEMO. AIP Conference Proceedings. 2020;**2254**: 020001. DOI: 10.1063/5.0014254

[70] Van Eester D, Lerche E. Integro-differential modeling of ICRH wave

propagation and damping at arbitrary cyclotron harmonics and wavelengths in tokamaks. AIP Conference Proceedings. 2014;**1580**:298. DOI: 10.1063/1.4864547

[71] Dumont RJ. Variational approach to radiofrequency waves in magnetic fusion devices. Nuclear Fusion. 2009;**49**:075033. DOI: 10.1088/0029-5515/49/7/075033

[72] Brambilla M. Numerical simulation of ion cyclotron waves in tokamak plasmas. Plasma Physics and Controlled Fusion. 1999;**41**:1. DOI: 10.1088/0741-3335/41/1/002

[73] Bilato R, Volpe F, Poli E, Köhn A, et al. Electron Cyclotron Heating in RFP plasmas. AIP Conference Proceedings. 2009;**1187**:81. DOI: 10.1063/1.3273812

[74] Kazakov YO, Ongena J, Wright JC, et al. Physics and applications of three-ion ICRF scenarios for fusion research. Physics of Plasmas. 2021;**28**:020501. DOI: 10.1063/5.0021818

[75] Lin Y, Wright J, Wukitch S. Physics basis for the ICRF system of the SPARC tokamak. Journal of Plasma Physics. 2020;**86**(5):865860506. DOI: 10.1017/S0022377820001269

[76] Lerche E, Goniche M, Jacquet P, Van Eester D, Bobkov V, Colas L, et al. Optimization of ICRH for core impurity control in JET-ILW. Nuclear Fusion. 2016;**56**:036022. DOI: 10.1088/0029-5515/56/3/036022

[77] TFR Group. Impurity pump-out at the two-ion hybrid resonance during ICRF experiments in TFR tokamak plasmas. Nuclear Fusion. 1982;**22**:956

[78] Rice JE, Lin Y, Perks CJ, Reinke ML, et al. Argon pumpout by ICRF waves in C-Mod L- and I-mode plasmas. Nuclear Fusion. 62 086009; 2022. DOI: 10.1088/1741-4326/ac6ef0

[79] Wukitch S et al. Towards ICRF antennas compatible with high performance plasmas: characterization and mitigation of ICRF antenna – plasma edge interaction. In: 22nd RFPPC Conference. France: Aix-en-Provence; 2017

[80] Hillairet J. Recent progress in ion cyclotron range of frequency in magnetic confinement fusion. From modelling technology and experiments. Reviews of Modern Physics. 2022. (Under reviewing process)

Chapter 8

Charged Particle Beam Injection into Magnetically Confined Plasmas

Wonyong Chung, Andi Tan and Christopher Tully

Abstract

As the principles underpinning magnetic confinement are contrary to allowing significant heat flow via charged particles into or out of a magnetically confined plasma, the approach of charged particle beam injection has been largely overlooked. The method of magnetic orbital angular momentum beam acceleration, developed by the PTOLEMY experiment, provides a new avenue for injecting charged particle beams into high magnetic field regions. Initial simulations show how this novel acceleration method can yield charge, mass, and heat flow into toroidal magnetic fields with important implications for fusion energy science. This chapter will review this new method in the context of charged particle beam injection methods and the relevance of these tools for plasma and fusion science.

Keywords: charged particle beam injection, heating, magnetic confinement, magnetic gradient drift, ITER

1. Introduction

The dream of harnessing energy from controlled nuclear fusion has been proposed for several decades. Intensifying climate change issues increase the desire for a clean and safe energy source. A fusion reactor based on magnetic confinement provides a promising configuration for controlled thermonuclear fusion. To fuse nuclei with large densities for an extended period, it is necessary to heat the plasma to overcome the Coulomb repulsion. The power ratio, Q, of the fusion output power to the input power is proportional to the fusion product $nT\tau_E$, where n and T are the central ion density and temperature [1]. The parameter τ_E is the energy confinement time. In December 2021, the Joint European Torus (JET) achieved a new record and produced 59 MJ of energy with a Q of 0.33 over a τ_E of 5 s [2]. Although remarkable progress has been made to achieve the required n, T, and τ_E, they have not been achieved in the same reactor configuration simultaneously.

To achieve an ignition condition where self-sustaining fusion is possible, additional energy-efficient heating is required. Ohmic heating from the toroidal current wanes at high temperatures. Two external sources are typically used to provide heating power, the resonant absorption of radio frequency electromagnetic waves and the injection of energetic neutral particle beams.

The injected beams are neutralized to prevent reflection due to the magnetic field. The neutralization process introduces inefficiency and complicates the instrumentation.

Alternatives to neutral particle beam injection, typically for non-equilibrium fusion reactors, have been explored using different acceleration technologies [3, 4]. The challenges of energy efficiency in particle acceleration are formidable given the high fraction of input power needed to operate relatively low-Q fusion reactors. Radio-frequency acceleration cavities and time-varying electromagnetic fields are, in general, prone to internal ohmic losses and self-heating. Static accelerating fields avoid the bulk of these losses, but are suited primarily for charged particle beams. By construction, the insertion and extraction of charged particles from magnetic confinement systems is thwarted except when necessary, as in the case of divertors. However, non-confining trajectories can be constructed under special conditions through the same processes of cyclotron orbit drift that plague steady-state operation.

In the transverse drift electromagnetic filter developed for the PTOLEMY experiment (Princeton Tritium Observatory for Light, Early-Universe, Massive-Neutrino Yield), a compact configuration of electromagnetic fields simultaneously transports and decelerates energetic electrons from the tritium β-decay endpoint starting in high magnetic fields of several Tesla to regions where both the kinetic energy and magnetic fields are reduced by several orders of magnitude [5, 6]. A new method is devised to accelerate low-energy charged particles into a high magnetic field region by operating the PTOLEMY filter in "reverse." This chapter presents the principles of this acceleration method and describes a possible application using the diagnostic port of the International Thermonuclear Experimental Reactor (ITER) [7].

2. Basics of charged particle beam injection

In this chapter, we use the convention that non-bolded symbols of vector quantities refer to the total magnitude unless a component is specified. The equation of motion of a charged particle of mass m and charge q in a magnetic field B is given by

$$\frac{\mathrm{d}}{\mathrm{d}t}\left(m\frac{\mathrm{d}\mathbf{r}}{\mathrm{d}t}\right) = q\frac{\mathrm{d}\mathbf{r}}{\mathrm{d}t} \times \mathbf{B}. \tag{1}$$

The Lorentz force on the right-hand side is perpendicular to the particle's velocity. In a uniform magnetic field, the particle's motion projected on a plane perpendicular to the magnetic field is circular, with a gyroradius given by

$$\rho = \frac{mv_\perp}{qB} = \frac{\sqrt{2mT_\perp}}{qB}, \tag{2}$$

with $T_\perp$ the transverse kinetic energy.

For a 1 MeV deuterium ion in a 5 T magnetic field, the gyroradius is about 0.04 m, a small fraction of a typical reactor radius. The ion beam injection energies must be relativistic to be commensurate with the reactor radius.

Relativistic ion beam injection introduces a number of inefficiencies. The plasma does not have the density required to stop energetic ions in a single transit, delivering limited power to the plasma and creating destructive irradiation of the reactor walls. The acceleration methods for relativistic beams involve time-varying fields that have several sources of intrinsic power loss.

In the following sections, charged particle injection of non-relativistic ions is re-examined as a transport mechanism that drifts charged ions from outside of the reactor volume to the surface of the plasma.

3. Magnetic orbital angular momentum beam acceleration

An alternative method to inject a charged particle beam is to create a beam of particles whose gyroradius is small compared to the transverse dimensions of the injection aperture. The particles are in cyclotron motion in a magnetic field that is relatively strong compared to their momentum. The acceleration mechanism stems from the ability of particles traveling in cyclotron motion in magnetic field gradients to do work. One, therefore, configures a magnetic geometry such that there is a transverse gradient along the average path of the beam. A complementary electric field is used to balance the gradient-*B* drift transverse to the average path of the beam and to accelerate the particles under the work of the magnetic field gradient. The acceleration process will be shown to be adiabatic for relevant injection energies and to maintain the magnetic moment invariance to a good accuracy after an initial stage of zero field ion source injection. The acceleration process does not affect the average linear momentum component of the beam. The increase in the charged particle kinetic energy follows from an increase in the magnetic orbital angular momentum.

3.1 Guiding-center drifts in adiabatic field conditions

When a charged particle gyrates in a magnetic field with a transverse gradient, the cyclotron-orbit averaged Guiding Center System (GCS) [8] motion can be described in terms of the drift terms of the virtual guiding-center particle if the spatial and temporal field variations within a single cyclotron orbit are taken to be adiabatic, i.e.,

$$\rho_c \ll \left|\frac{B}{\nabla B}\right|, \left|\frac{E}{\nabla E}\right|; \text{and} \tag{3}$$

$$\tau_c \ll \left|\frac{B}{dB/dt}\right|, \left|\frac{E}{dE/dt}\right|; \tag{4}$$

where ρ_c is the Larmor radius and τ_c the cyclotron period. Under the conditions specified by Eq. (3) and (4), the first adiabatic invariant μ,

$$\mu = \frac{mv_\perp^2}{2B} = \frac{T_\perp}{B}, \tag{5}$$

accurately describes an invariant quantity preserved in the motion of the particle [9, 10] and shows that an increase in the magnetic field magnitude is accompanied by a proportional increase in the transverse kinetic energy. Additionally, the deviation of the GCS trajectory from the direction of the magnetic field lines can be described in terms of four fundamental drift terms,

$$\boldsymbol{V}_D = \boldsymbol{V}_\perp = \left(q\boldsymbol{E} + \boldsymbol{F} - \mu\nabla B - m\frac{d\boldsymbol{V}}{dt}\right) \times \frac{\boldsymbol{B}}{qB^2}, \tag{6}$$

where $V_\perp$ is the perpendicular component of the GCS velocity with respect to the magnetic field line. The transverse drift velocity, V_D, is composed of individual terms, as appear in Eq. (6) from left to right, known as (1) the $E \times B$ drift; (2) the external force drift; (3) the gradient-B drift; and (4) the inertial drift [8]. It is possible to configure the electric and magnetic field parameters to manipulate certain drift terms to produce a net linear trajectory in the transverse direction [5].

3.2 Drifts and work

The gradient-B drift is able to drive a charged particle up or down an electrostatic potential. This ability to do work, at first, seems contrary to the notion that magnetic fields do not do work on charged particles, as seen in Eq. (1), from the cross-product. Similarly, under the motion of $E \times B$ drift alone, the cross-product bars work as the electrons will drift on surfaces of constant voltage. This can also be understood by considering that it is always possible to boost into a frame in which the $E \times B$ drift is zero.

In contrast, a gradient-B drift due to a spatially varying magnetic field implies a time-varying electric field that cannot be boosted to zero. By itself, i.e., with a magnetic field and no electric field, a gradient-B does no work because there is nothing to do work against. However, when accompanied by an external $E \times B$ drift, the external electric potential provides a surface against which the gradient-B drift can do work on. The internal rotational kinetic energy of gyromotion of the virtual guiding-center particle is reduced for a corresponding increase in voltage potential. This is described by inserting terms from Eq. (6),

$$\frac{dT_\perp}{dt} = -q\boldsymbol{E} \cdot \boldsymbol{V}_D = -q\boldsymbol{E} \cdot (q\boldsymbol{E} - \mu \nabla B) \times \frac{\boldsymbol{B}}{qB^2} = \frac{\mu}{B^2} \boldsymbol{E} \cdot (\nabla B \times \boldsymbol{B}), \tag{7}$$

where $T_\perp$ is the internal kinetic energy of gyromotion in the GCS frame [8].

3.3 Balanced drift

To produce a filter or accelerator based on the drift terms in Eq. (6), the external force and inertial drift terms are first taken to be zero, leaving only the electric and gradient-B drifts to be configured such that the total net drift is along a straight line parallel to the direction of the magnetic field gradient. The gradient-B drift alone is orthogonal to the direction of the magnetic field gradient, so the first step is to create a component of the $E \times B$ drift that exactly counters the gradient-B drift. From Eq. (6), this specifies the requirement,

$$q\boldsymbol{E}_\parallel \times \boldsymbol{B} = \mu \nabla B \times \boldsymbol{B}, \tag{8}$$

where $E_\parallel$ is the component of electric field parallel to the magnetic field gradient. In general, the ratio of the parallel electric field to the magnitude of the magnetic field to meet this condition depends on the ratio μ/q times the fractional rate of change of the transverse component of the magnetic field along the direction of the magnetic field gradient. For an exponentially falling transverse field, the fractional rate of change is $1/\lambda$, the characteristic exponential length scale in units of transverse distance.

To introduce work, the electric field is tilted by adding an additional component, $E_\perp$, that is orthogonal to the direction of the magnetic field gradient. The $E_\perp \times B$ drift is what moves the charged particle either against or along the magnetic field gradient. As the components of $E_\parallel$ and $E_\perp$ are in vacuum, the relationship between the components follows from solving Maxwell's equations for a set of voltage plates above and below the direction of the balanced drift. Explicit solutions have been found [5]. Given that the magnitudes of $E_\parallel$ and $E_\perp$ are related, it is not surprising that the net drift velocity along the acceleration direction is constant. There is no linear momentum acceleration present. The acceleration occurs through the increase in the transverse kinetic energy component, the magnetic orbital angular momentum, during a process of constant drift along the magnetic field gradient.

3.4 Performance

Because the orbital magnetic moment $\mu = T_\perp/B$ is invariant, if the B field increases (or decreases) exponentially along the trajectory of the particle, so must its transverse kinetic energy. **Figure 1** shows the trajectory of a deuterium ion in a balanced drift with an initial kinetic energy of 20 keV at 0.2 T, going to a 4.7 T region with 1 MeV final kinetic energy. In this simulation, using CST studio [11], the magnetic field (**Figure 2**) is scaled from the one produced by the PTOLEMY magnet [6]. The maximum of the B_x field is set to 4.7 T at Z = 1.6 m to match the toroidal field near the interface of the upper port of ITER [12]. The electric field as in **Figure 3** is generated by a similar electrode structure as in the PTOLEMY transverse drift filter [6]. The dimensions are scaled up such that the distance between the electrodes is 0.3 m.

3.5 Injection

The net drift is along the direction of the magnetic field gradient and drives the guiding-center of the beam to cross equipotential lines and accelerates the particles. As the beam drifts in the direction of ∇B, it naturally reaches its maximum kinetic energy upon entering the toroidal magnet of a tokamak.

Via the foregoing mechanism, initial simulations of injecting deuterium ions indicate successful delivery of the beam, as shown in **Figure 4**. Once the particle leaves

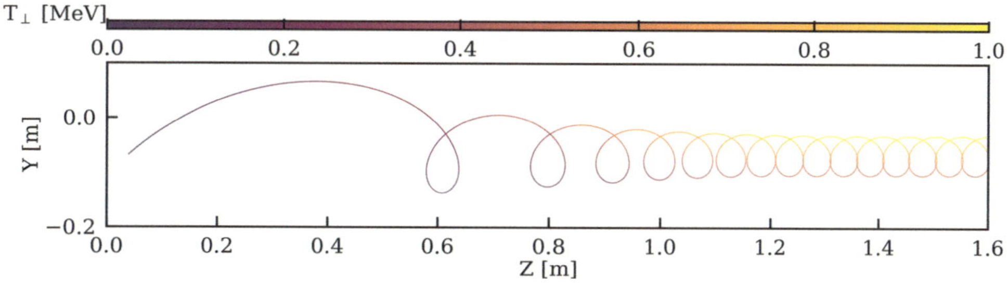

Figure 1.
The trajectory of a deuterium ion in a transverse drift accelerator is shown with the low energy ion source on the left at a low magnetic field region and the high energy ion exiting the accelerator on the right in a region of high magnetic field. The net vertical drift is balanced to zero by construction as the ion drifts at constant velocity from left to right while climbing the magnetic field gradient. The trajectory is computed using the CST software suite. The color scale indicates the kinetic energy of the ion increasing from 20 keV at 0.2 T to 1 MeV at 4.7 T.

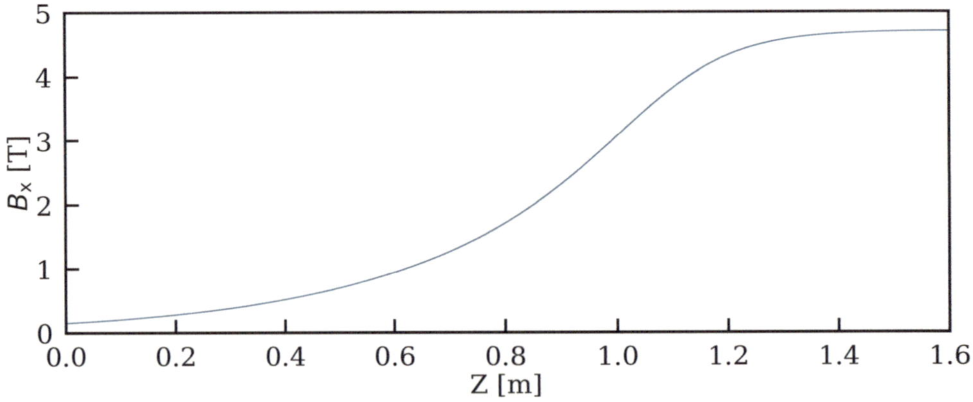

Figure 2.
*The profile of the transverse magnetic field in the CST simulation for the trajectory shown in **Figure 1**. It is derived from scaling the magnetic field produced by the PTOLEMY magnet to match a maximum B_x of 4.7 T.*

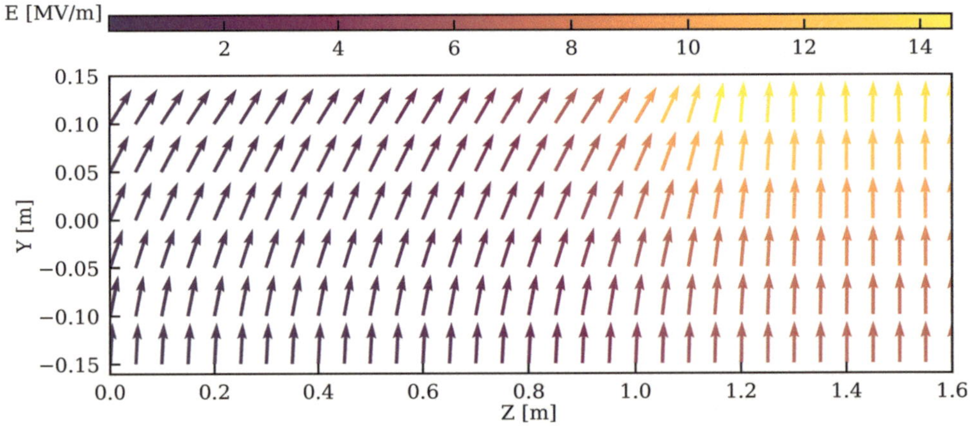

Figure 3.
*The electric field in the transverse plane in the drift region for the trajectory shown in **Figure 1**. The color (orientation) of the arrow indicates the magnitude (direction) of the field at its tail-end position.*

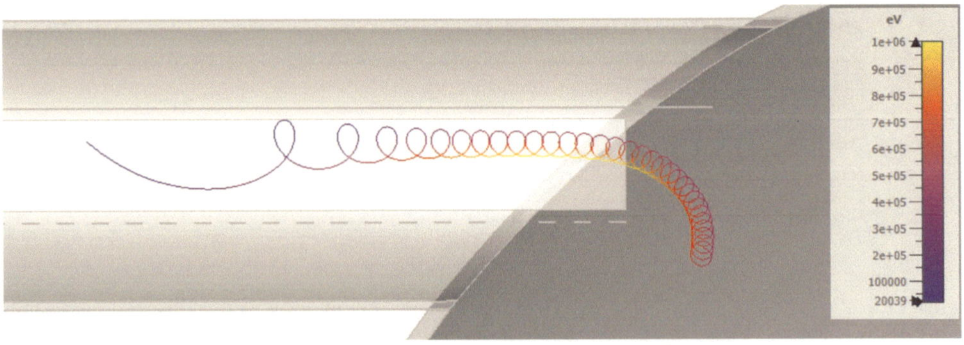

Figure 4.
*Simulation of the injection of deuterium ions into a 1/R magnetic field in a tokamak using the accelerating structure of **Figure 1**. Upon exiting the accelerator, the energetic ion continues to drift toward the plasma confinement region under the 1/R magnetic field gradient-B drift of the tokamak.*

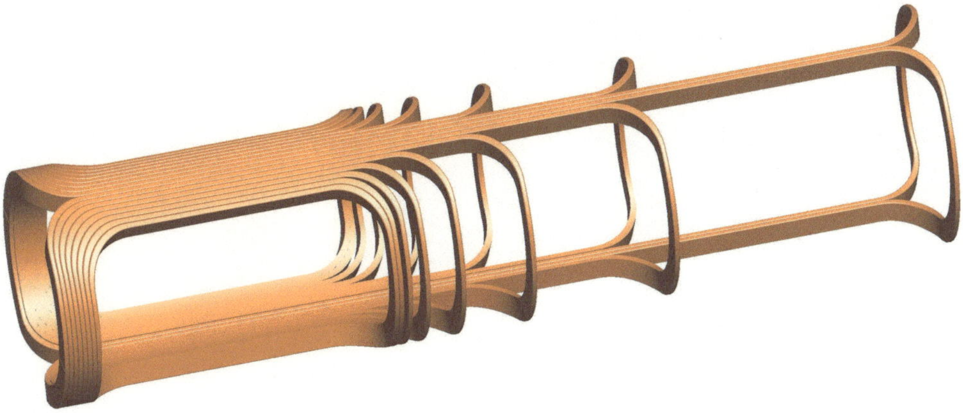

Figure 5.
A conceptual design of a tapered dipole magnet winding to generate the desired magnetic field for magnetic orbital angular momentum beam acceleration. The winding of the coils follows the surface of a cylindrical vacuum insertion port, similar to dipole magnets used in circular proton beam accelerators.

the injection port, the gradient of the $1/R$ toroidal magnetic field drifts the ions into the center of the plasma. The relatively hot thermal temperature of the 1 MeV deuterium ions will thermalize through Coulomb interactions with the plasma. The injection mechanism supports a range of injection energies and ion species. For instance, injection of 4 MeV α-particles through the ITER diagnostic port may be an effective way of studying the effects of fusion final-state ion interactions on the plasma. Charge neutralization can be achieved by instrumenting ion (electron) injection ports on the top (bottom) of the tokamak. The gradient-B drift will drift ions downward (for a given orientation of the azimuthal toroidal magnetic field) and electrons (or negative ions) upward.

The desired injection magnetic field, as described in Ref. [6], can be produced by a tapered dipole magnet with a superconductor winding, as in **Figure 5**.

A field cage with a number of electrodes can be placed inside the magnet to produce the corresponding electric field. Such a magnet is compact and can be placed within a counter-dipole coil in the upper diagnostic ports of ITER, as shown in **Figure 6**. The counter-dipole creates a zero field region for the ion source and reduces Lorentz forces on the primary reactor coils. The details of this magnet and the field cage are beyond the scope of this chapter.

4. Energy efficiency

An important aspect of magnetic orbital angular momentum acceleration for fusion energy efficiency is the reliance solely on static electric and magnetic fields. The power loading during injection on the accelerating plate voltages draws from highly efficient DC power supplies. Above all, the largest inefficiency of neutral beam injection, the neutralization, is avoided with direct charged particle injection. The high currents and efficient production of positive ions saves on power losses at the source relative to the negative ion beams used for neutral beam injection [13]. The inefficiencies and beam energy limitations associated with neutralization and

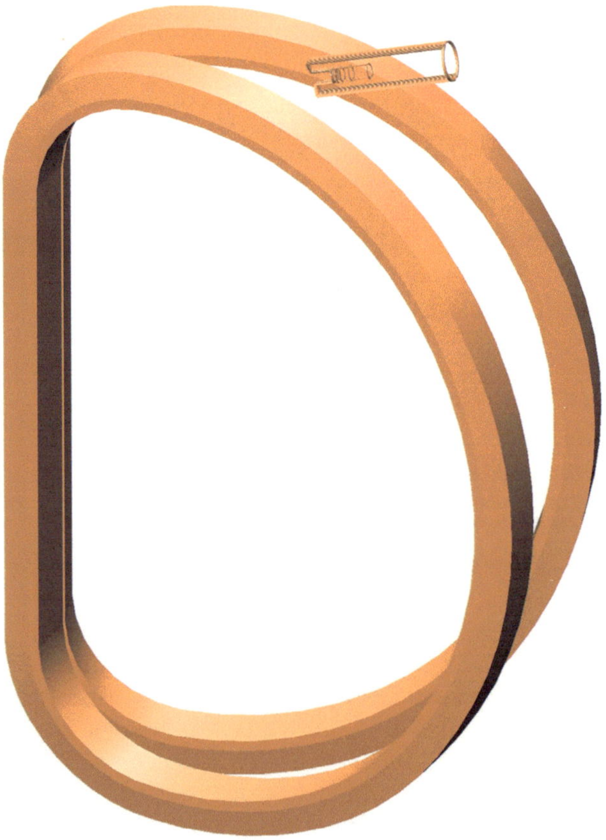

Figure 6.
*A tapered dipole magnet, as shown in **Figure** 5, within a counter-dipole winding placed in the upper port between two adjacent ITER toroidal field coils. The counter-dipole reduces the magnetic forces between the injection system and the tokamak field windings.*

neutral beam injection introduce approximately a factor of 2 loss in absolute power efficiency [14].

5. Conclusions

Fusion reactor science is on the brink of a major advance toward sustained clean energy reactors. Reducing inefficiency in the particle beam injection systems is a promising direction toward achieving ignition conditions without compromising reactor operation. A surprising and yet potentially revolutionary approach to improving beam heating is through a new particle acceleration method called magnetic orbital angular momentum beam acceleration. With this technique, charged particle beam injection into magnetically confined plasmas becomes possible. The relevant parameters for charged particle beam injection are presented with simulated geometries demonstrating the feasibility of implementing this system with the diagnostic port of ITER. Charged particle beam injection provides a new tool for fusion reactors to deliver charge, mass, and heat flow into the plasma. The large gain in energy efficiency for charged particle injection is the most advantageous factor in comparing with neutral beam injection.

Acknowledgements

This research was supported by the Simons Foundation (#377485) and the John Templeton Foundation (#61814).

Author details

Wonyong Chung, Andi Tan* and Christopher Tully
Department of Physics, Princeton University, Princeton, New Jersey, USA

*Address all correspondence to: andit@princeton.edu

References

[1] Wesson J, Campbell DJ. Tokamaks. Oxford, England: Oxford University Press; 2011. Available online: https://global.oup.com/academic/product/tokamaks-9780199592234

[2] Gibney E. Nuclear-fusion reactor smashes energy record. Nature. 2022; **602**(7897):371-371

[3] Liu KF, Chao AW. Accelerator based fusion reactor. Nuclear Fusion. 2017; **57**(8):084002

[4] Labaune C, Baccou C, Depierreux S, Goyon C, Loisel G, Yahia V, et al. Fusion reactions initiated by laser-accelerated particle beams in a laser-produced plasma. Nature Communications. 2013; **4**(1):1-6

[5] Betti MG, Biasotti M, Bosca A, Calle F, Carabe-Lopez J, Cavoto G, et al. A design for an electromagnetic filter for precision energy measurements at the tritium endpoint. Progress in Particle and Nuclear Physics. 2019;**106**:120-131

[6] Apponi A, Betti MG, Borghesi M, Canci N, Cavoto G, Chang C, et al. Implementation and optimization of the PTOLEMY transverse drift electromagnetic filter. Journal of Instrumentation. 2022;**17**(05):P05021. DOI: 10.1088/1748-0221/17/05/p05021

[7] International Thermonuclear Experimental Reactor (ITER). 2022. Available from: http://www.iter.org [Accessed: 2022-06]

[8] Roederer J, Zhang H. Particle fluxes, distribution functions and violation of invariants. In: Dynamics of Magnetically Trapped Particles. Berlin, Heidelberg: Springer-Verlag; 2014. pp. 89-122. Available online: https://link.springer.com/book/10.1007/978-3-642-41530-2

[9] Alfvén H. On the motion of a charged particle in a magnetic field. Arkiv foer Matematik, Astronomi, och Fysik. 1940; **25B**(1–20):29

[10] Cary JR, Brizard AJ. Hamiltonian theory of guiding-center motion. Reviews of Modern Physics. 2009;**81**(2): 693

[11] Dassault Systèmes. CST Studio Suite; Available from: http://www.cst.com

[12] International Atomic Energy Agency. Summary of the ITER Final Design Report. No. 22 in ITER EDA Documentation Series. Vienna: International Atomic Energy Agency; 2001. Available from: https: //www.iaea.org/publications/6442/ summary-of-the-iter-final-design-report

[13] Hemsworth RS, Inoue T. Positive and negative ion sources for magnetic fusion. IEEE Transactions on Plasma Science. 2005;**33**(6):1799-1813

[14] Hopf C, Starnella G, den Harder N, Fantz U. Neutral beam injection for fusion reactors: Technological constraints versus functional requirements. Nuclear Fusion. 2021; **61**(10):106032. DOI: 10.1088/1741-4326/ac227a

Chapter 9

Decoupling Techniques for Coupled PDE Models in Fluid Dynamics

Mingchao Cai, Mo Mu and Lian Zhang

Abstract

We review decoupling techniques for coupled PDE models in fluid dynamics. In particular, we are interested in the coupled models for fluid flow interacting with porous media flow and the fluid structure interaction (FSI) models. For coupled models for fluid flow interacting with porous media flow, we present decoupled preconditioning techniques, two-level and multilevel methods, Newton-type linearization-based two-level and multilevel algorithms, and partitioned time-stepping methods. The main theory and some numerical experiments are given to illustrate the effectiveness and efficiency of these methods. For the FSI models, partitioned time-stepping algorithms and a multirate time-stepping algorithm are carefully studied and analyzed. Numerical experiments are presented to highlight the advantages of these methods.

Keywords: decoupling, linearization, Stokes/Darcy model, FSI model, finite element, two-level method, Robin-Neumann scheme, β-scheme

1. Introduction

Coupled PDE models have wide applications in the real world. For example, in fluid dynamics, there are two-phase flow models, fluid structure interaction models, heat transfer models in fluids etc. In this work, we focus on two typical models: coupled models for describing fluid flow interacting with porous media flow, and fluid structure interaction (FSI) models. The first type of models have been validated by experiments [1] and then justified by using homogenization theory [2, 3]. Applications include the environmental engineering problem of groundwater contamination through rivers and the geoscience problem of surface flows filtrating in vuggy porous media. The second type of models come from many practical applications. For example, blood flow interacting with vessel wall, compressible fluids interacting with aircraft wings, as well as slamming and whipping response of ship structure to water flow. These models are typical multidomain coupled PDE models with multiphysics. Due to the heterogenenities in subdomain models, it is very difficult to find a unified approach to solve the different subdomain models simultaneously. Moreover, some coupled models have nonlinearity in either subdomain problems or interface coupling terms. To deal with the difficulties caused by the coupling of different submodels and the nonlinearity [4–8], we discuss some decoupling techniques [4, 9–20], which have shown to be very effective and efficient. Among them, we will emphasize the work proposed by our group members and highlight the novelty and importance of these algorithms.

The rest of the paper is organized as follows. In Section 2, we introduce the coupled fluid flow/porous media interacting models. Some decoupling techniques, specifically, decoupled preconditioners, decoupled two-level and multi-level methods, and partitioned time schemes will be presented and analyzed. In Section 3, we present fluid structure interaction models. Partitioned decoupling algorithms include the Robin-Neumann scheme [14, 15], the β-scheme [13, 21], and the multirate partitioned schemes [22, 23] will be briefly introduced. Concluding remarks are drawn in Section 4.

2. Decoupled algorithms for the coupled models of fluid flow interacting with porous flow

For the linear cases of the coupled models for fluid flow interacting with porous media flow, we refer to the Stokes/Darcy model studied in [7, 11, 18, 20, 24–30]. For the nonlinear case, we refer to the coupled nonlinear Navier–Stokes/Darcy model [4–6].

2.1 Coupled models for fluid flow interacting with porous media flow

Let $\Omega \subset R^d$ be a domain consisting of a fluid region Ω_f and a porous media region Ω_p separated by an interface Γ, as shown in **Figure 1**, where $d = 2$ or 3, $\Omega = \Omega_f \cup \Omega_p$ and $\Gamma = \overline{\Omega}_f \cap \overline{\Omega}_p$. Let $\boldsymbol{n}_f$ and $\boldsymbol{n}_p$ denote the unit outward normal directions on $\partial\Omega_f$ and $\partial\Omega_p$. The interface Γ is assumed to be smooth enough as in [6].

For incompressible Newtonian fluid flow, Navier–Stokes equations of the stress-divergence form are usually used [31, 32]. $\forall t \geq 0$, $\forall \boldsymbol{x} \in \Omega_f$,

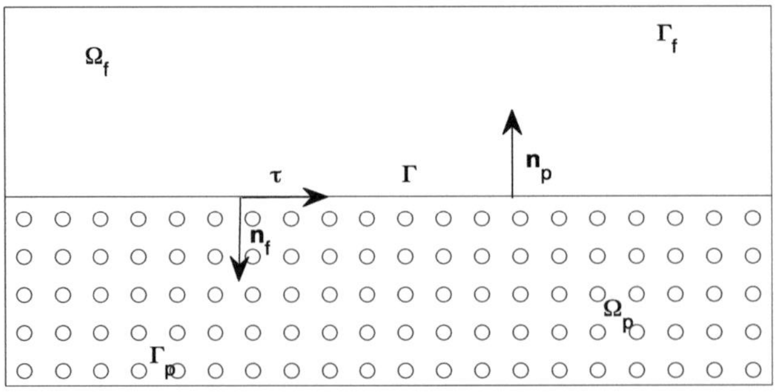

Figure 1.
A global domain Ω consisting of a fluid region Ω_f and a porous media region Ω_p separated by an interface Γ.

$$\begin{cases} \rho_f \left(\frac{\partial \boldsymbol{u}}{\partial t} + (\boldsymbol{u} \cdot \nabla) \boldsymbol{u} \right) - div\, \boldsymbol{T}(\boldsymbol{u}, p) = \boldsymbol{f}_f, \\ div\, \boldsymbol{u} = 0, \end{cases} \tag{1}$$

where ρ_f is the fluid density, $\boldsymbol{u}$ is the velocity vector, p is the pressure, $\boldsymbol{f}_f$ is the external force,

$$\boldsymbol{T}(\boldsymbol{u}, p) = 2\nu \boldsymbol{D}(\boldsymbol{u}) - p\boldsymbol{I}, \quad \text{with} \quad \boldsymbol{D}(\boldsymbol{u}) = \frac{1}{2}\left[\nabla \boldsymbol{u} + (\nabla \boldsymbol{u})^T\right], \tag{2}$$

is the stress tensor with $\nu > 0$ being the kinematic viscosity. By dropping the term $\frac{\partial \boldsymbol{u}}{\partial t}$ in (1), the steady state Navier–Stokes equations read as: $\forall \boldsymbol{x} \in \Omega_f$,

$$\begin{cases} \rho_f (\boldsymbol{u} \cdot \nabla) \boldsymbol{u} - \operatorname{div} \boldsymbol{T}(\boldsymbol{u}, p) = \boldsymbol{f}_f, \\ \operatorname{div} \boldsymbol{u} = 0. \end{cases} \tag{3}$$

In strong form,

$$-\operatorname{div} \boldsymbol{T}(\boldsymbol{u}, p) = -\nu \Delta \boldsymbol{u} + \nabla p, \tag{4}$$

because the fluid flow is assumed to be divergence free and $\operatorname{div}\left(\nabla^T \boldsymbol{u}\right) = 0$ holds.

Among various porous media flow models, Darcy's law is the most favored. The governing variable in Ω_p is the so-called *piezometric head* or *pressure head*,

$$\phi = z + \frac{p_p}{\rho_f g}. \tag{5}$$

Here, z is the elevation from a reference level (for simplicity, z is assumed to be 0). Darcy's law states that the velocity $\boldsymbol{u}_p$ (also called seepage velocity) in the porous media region is proportional to the gradient of ϕ [27, 33].

$$\boldsymbol{u}_p = -\boldsymbol{K} \nabla \phi. \tag{6}$$

We assume that

$$\alpha_1(\boldsymbol{x}, \boldsymbol{x}) \le (\boldsymbol{K}\boldsymbol{x}, \boldsymbol{x}) \le \alpha_2(\boldsymbol{x}, \boldsymbol{x}), \quad \forall \boldsymbol{x} \in \Omega_p. \tag{7}$$

Moreover, the divergence of the seepage velocity equals to the source term. This leads to the following steady state equation:

$$-\operatorname{div}(\boldsymbol{K} \nabla \phi) = f_p. \tag{8}$$

In the time-dependent case, the governing equations in Ω_p reads as:

$$S_0 \phi_t - \operatorname{div}(\boldsymbol{K} \nabla \phi) = f_p, \tag{9}$$

where S_0 is a specific storage and f_p is a source term.

No matter time-dependent or steady state, the key part of the coupled model is a set of interface conditions, which describe the interaction mechanism of the two different types of flows. The following interface conditions have been extensively used and studied in the literature [1–3, 7, 27, 28]:

$$\begin{cases} \boldsymbol{u}\cdot\boldsymbol{n}_f = \boldsymbol{u}_p\cdot\boldsymbol{n}_f = -\boldsymbol{K}\nabla\phi\cdot\boldsymbol{n}_f, \\ -\nu(\nabla\boldsymbol{u}\boldsymbol{n}_f)\cdot\boldsymbol{n}_f + p = \rho g\phi, \\ -\nu(\nabla\boldsymbol{u}\boldsymbol{n}_f)\cdot\tau_i = \dfrac{\nu\alpha_{BJS}}{\sqrt{\nu\tau_i\cdot\boldsymbol{K}\tau_i}}\boldsymbol{u}\cdot\tau_i, \quad i = 1, \dots, d-1. \end{cases} \tag{10}$$

Here, $\{\tau_i\}_{i=1}^{d-1}$ is the unit tangent vector on Γ, α_{BJS} is a positive parameter depending on the properties of the porous medium. The first interface condition ensures mass conservation across Γ. The second one is the balance of normal forces across the interface. The third condition is well known as Beavers-Joseph-Saffman's law [1, 2], which states that the slip velocity is proportional to the shear stress along Γ. Without loss of generality, we impose homogeneous Dirichlet boundary conditions on both of the external boundaries:

$$\begin{cases} \boldsymbol{u} = \boldsymbol{0} \quad \text{on} \quad \Gamma_f, \\ \phi = 0 \quad \text{on} \quad \Gamma_p. \end{cases} \tag{11}$$

The proper functional spaces for $\boldsymbol{u}$, p and ϕ are

$$\begin{aligned} \boldsymbol{X}_f &= \left\{\boldsymbol{v}\in\boldsymbol{H}^1(\Omega_f) = \left(H^1(\Omega_f)\right)^d \middle|\ \boldsymbol{v} = \boldsymbol{0} \text{ on } \Gamma_f\right\}, \\ Q = L^2(\Omega_f), &\quad X_p = \{\psi\in H^1(\Omega_p) |\ \psi = 0 \text{ on } \Gamma_p\}. \end{aligned} \tag{12}$$

Moreover, we denote $\underline{X} = \boldsymbol{X}_f \times X_p$ for ease of presentation. Multiplying test functions to (3) and (8), integrating by parts and plugging in the interface boundary conditions (10)–(11), we have the weak form of the coupled Navier–Stokes/Darcy model: find $\underline{u} = (\boldsymbol{u},\phi)\in\underline{X}, p\in Q$ such that

$$\begin{cases} a(\underline{u},\underline{v}) + c(\boldsymbol{u},\boldsymbol{u},\boldsymbol{v}) + b(\boldsymbol{v},p) = f(\underline{v}) & \forall\underline{v} = (\boldsymbol{v},\psi)\in\underline{X}, \\ b(\boldsymbol{u},q) = 0 & \forall q\in Q, \end{cases} \tag{13}$$

where

$$\begin{aligned} a(\underline{u},\underline{v}) &= a_f(\boldsymbol{u},\boldsymbol{v}) + a_p(\phi,\psi) + a_\Gamma(\underline{u},\underline{v}), \quad b(\boldsymbol{v},p) = -\int_{\Omega_f} p\nabla\cdot\boldsymbol{v}, \\ c(\boldsymbol{u},\boldsymbol{v},\boldsymbol{w}) &= \rho\int_{\Omega_f}(\boldsymbol{u}\cdot\nabla)\boldsymbol{v}\cdot\boldsymbol{w}, \quad f(\underline{v}) = \int_{\Omega_f}\boldsymbol{f}_f\cdot\boldsymbol{v} + \rho g\int_{\Omega_p} f_p\psi \end{aligned} \tag{14}$$

with

$$\begin{aligned} a_f(\boldsymbol{u},\boldsymbol{v}) &= \nu\int_{\Omega_f}\nabla\boldsymbol{u} : \nabla\boldsymbol{v} + \sum_{i=1}^{d-1}\frac{\nu\alpha_{BJS}}{\sqrt{\nu\tau_i\cdot\boldsymbol{K}\tau_i}}\int_\Gamma(\boldsymbol{u}\cdot\tau_i)(\boldsymbol{v}\cdot\tau_i), \\ a_p(\phi,\psi) &= \rho g\int_{\Omega_p}\nabla\psi\cdot\boldsymbol{K}\nabla\phi, \quad a_\Gamma(\underline{u},\underline{v}) = \rho g\int_\Gamma(\phi\boldsymbol{v} - \psi\boldsymbol{u})\cdot\boldsymbol{n}_f. \end{aligned} \tag{15}$$

For the wellposedness of the coupled Navier–Stokes/Darcy model, we refer to the recent results in [5, 6, 34]. It is shown in [34] that if the viscosity is sufficiently large and the normal velocity across the interface is sufficiently small, then the problem (14) is wellposed. We follow the assumptions in [34]. In addition to these assumptions on model parameters and variables, we shall frequently use these properties: $a(\cdot,\cdot)$ is bounded and coercive; $b(\cdot,\cdot)$ is bounded and satisfies the inf-sup condition [27, 35]; and the nonlinear term can be bounded by using the H^1 norm of the three components [31, 36].

We partition Ω_f and Ω_p by quasi-uniform triangulations $\mathcal{T}_{f,h}$ and $\mathcal{T}_{p,h}$ with a characteristic meshsize h. Here, we require that the two subdomain triangulations coincide at Γ. The corresponding finite element spaces are denoted by $\boldsymbol{X}_{f,h} \times Q_h \subset \boldsymbol{X}_f \times Q$ and $X_{p,h} \subset X_p$, respectively. Moreover, $\boldsymbol{X}_{f,h} \times Q_h$ needs to be stable, i.e., there exists a positive constant β such that

$$\sup_{\boldsymbol{v}_h \in \boldsymbol{X}_{f,h}} \frac{b(\boldsymbol{v}_h, q_h)}{|\boldsymbol{v}_h|_{1,\Omega_f}} \geq \beta \|q_h\|_{0,\Omega_f} \qquad \forall q_h \in Q_h. \tag{16}$$

We assume that the solution of (13) is smooth enough and the finite element spaces have the following typical approximation properties: for all $(\boldsymbol{u}, p) \in \boldsymbol{H}^{k+1}(\Omega_f) \cap \boldsymbol{X}_f \times H^k(\Omega_f)$ and $\phi \in H^{k+1}(\Omega_p) \cap X_p$,

$$\inf_{\boldsymbol{v}_h \in \boldsymbol{X}_{f,h}, q_h \in Q_h} \left\{ h|\boldsymbol{u} - \boldsymbol{v}_h|_{1,\Omega_f} + \|\boldsymbol{u} - \boldsymbol{v}_h\|_{0,\Omega_f} + h\|p - q_h\|_{0,\Omega_f} \right\} \lesssim h^{k+1}\left(|\boldsymbol{u}|_{k+1,\Omega_f} + |p|_{k,\Omega_f} \right), \tag{17}$$

$$\inf_{\psi_h \in X_{p,h}} \left\{ h|\phi - \psi_h|_{1,\Omega_p} + \|\phi - \psi_h\|_{0,\Omega_p} \right\} \lesssim h^{k+1} |\phi|_{k+1,\Omega_p}. \tag{18}$$

To satisfy the discrete inf-sup condition and the approximation properties (17)–(18), if $k = 1$, one may apply the Mini elements [31, 37] in Ω_f and the piecewise linear elements in Ω_p; if $k \geq 2$, the k-th order Taylor-Hood elements [31, 37, 38] can be applied in Ω_f and the piecewise k-th order elements can be adopted in Ω_p.

Coupled Algorithm: A conventional finite element discretization applied to the model problem (13) leads to the discrete problem: Find $\underline{u}_h = (\boldsymbol{u}_h, \phi_h) \in \underline{X}_h = \boldsymbol{X}_{f,h} \times X_{p,h}, p_h \in Q_h$ such that

$$\begin{cases} a(\underline{u}_h, \underline{v}_h) + c(\boldsymbol{u}_h, \boldsymbol{u}_h, \boldsymbol{v}_h) + b(\boldsymbol{v}_h, p_h) = f(\underline{v}_h) & \forall \underline{v}_h = (\boldsymbol{v}_h, \psi_h) \in \underline{X}_h, \\ b(\boldsymbol{u}_h, q_h) = 0, \quad \forall q_h \in Q_h. \end{cases} \tag{19}$$

2.2 Decoupled algorithms in the preconditioning steps

As an illustration of decoupled preconditioning techniques, we will consider the linear Stokes/Darcy model, whose weak form is (19) while the nonlinear term is dropped. We note that the discrete model in the operator form is

$$\begin{bmatrix} A_p & A_\Gamma^T & \boldsymbol{0} \\ -A_\Gamma & A_f & B_f^T \\ \boldsymbol{0} & B_f & \boldsymbol{0} \end{bmatrix} \begin{bmatrix} \phi_h \\ \boldsymbol{u}_h \\ p_h \end{bmatrix} = \begin{bmatrix} \boldsymbol{f}_{f,h} \\ \boldsymbol{f}_{p,h} \\ g_h \end{bmatrix}. \tag{20}$$

Here A_f, A_p, A_Γ, and $B = (\mathbf{0}, B_f)$ are the corresponding linear operators induced by the corresponding bilinear forms in (15) and (16). We denote

$$A = \begin{bmatrix} A_p & A_\Gamma^T \\ -A_\Gamma & A_f \end{bmatrix}, \quad \text{and} \quad M = \begin{bmatrix} \boldsymbol{A} & B^T \\ B & \mathbf{0} \end{bmatrix}. \tag{21}$$

By discarding the coupling interface terms and plugging in $\frac{1}{\nu}\boldsymbol{I}$ (which leads to pressure mass matrix) in the (2, 2) block, we have the following block-diagonal decoupled preconditioner:

$$P_M = \begin{bmatrix} \boldsymbol{A}_0 & \mathbf{0} \\ \mathbf{0} & \frac{1}{\nu}\boldsymbol{I}_h \end{bmatrix}, \quad \text{with} \quad A_0 = \begin{bmatrix} A_p & \mathbf{0} \\ \mathbf{0} & A_f \end{bmatrix}. \tag{22}$$

By keeping one of B and B^T, one can easily construct block-triangular decoupled preconditioner. GMRES method is used as the outer iterative method. Block diagonal or block-triangular preconditioners are used in the inner iteration. The effectiveness and the efficiency of the preconditioners have been verified in [9, 39, 40]. In the implementation, A_0^{-1} and the inverse for $\frac{1}{\nu}\boldsymbol{I}$ should be realized by applying a Multigrid algorithm or domain decomposition methods. Particularly, when Krylov subspace methods are used, these inverses should be applied inexactly (for example, using one V-cycle Multigrid algorithm to provide approximate inverses).

Let us denote

$$P_\pm = \begin{pmatrix} A_0 & 0 \\ 0 & \pm\frac{1}{\nu}\boldsymbol{I}_h \end{pmatrix}, \tag{23}$$

We also propose another preconditioner of the block triangular type:

$$P_{T_1} = \begin{pmatrix} A_0 & 0 \\ B & -\frac{1}{\nu}\boldsymbol{I}_h \end{pmatrix}, \tag{24}$$

by retaining the divergence operator. This preconditioner is still decoupled as the computation can be carried out in a block forward substitution manner. As an illustration, we illustrate the importance of using preconditioners in **Table 1**.

We use $**$ to indicate that the iteration does not converge within the prescribed maximum number of iterations. $N(P)$ refers to the number of iterations with a preconditioner P, and no preconditioner is applied when P is simply I. From the table, it is clear that both P_- and P_{T_1} accelerate the convergence of the GMRES method. The number of iterations based on the two preconditioners is independent of the mesh refinement. More numerical experiments for testing the robustness with respect to the physical parameters can be found in [9, 39].

2.3 Decoupling and linearization by two-level and multi-level algorithms

For the mixed Stokes/Darcy model, Mu and Xu in [11] propose a two-grid method in which the coarse grid solution is used to supplement the boundary conditions at the

h	DOF	$N(I)$	$N(P_-)$	$N(P_{T_1})$
2^{-2}	268	186	41	20
2^{-3}	948	432	45	22
2^{-4}	3556	* *	48	22
2^{-5}	13,764	* *	46	22
2^{-6}	54,148	* *	46	22

Table 1.
Number of iterations for the GMRES method without and with the two preconditioners P_- and P_{T_1}.

interface for both of the two subproblems. The **Two-grid Algorithm** proposed in [11] is composed by the following two steps.

1. Solve the linear part of problem $(2.7)_H$ on a coarse grid: find $\underline{u}_H = (\boldsymbol{u}_H, \phi_H) \in \underline{X}_H \subset \underline{X}_h, p_H \in Q_H \subset Q_h$ such that

$$\begin{cases} a(\underline{u}_H, \underline{v}_H) + b(\underline{v}_H, p_H) = (f, \underline{v}_H), & \forall \underline{v}_H = (\boldsymbol{v}_H, \psi_H) \in \underline{X}_H, \\ b(\underline{u}_H, q_H) = 0, & \forall q_H \in Q_H; \end{cases} \tag{25}$$

2. Solve a modified fine grid problem: find $\underline{u}^H = (\boldsymbol{u}^h, \phi^h) \in \underline{X}_h, p^h \in Q_h$ such that

$$\begin{cases} a(\underline{u}^H, v_h) + b(v_h, p^h) = (f, v_h) - a_\Gamma(\underline{u}_H, v_h), & \forall v_h \in \underline{X}_h, \\ b(\underline{u}^H, q_h) = 0, & \forall q_h \in Q_h. \end{cases} \tag{26}$$

The main theoretical results for the two grid algorithm are as follows.

Theorem 1. *Let $(\underline{u}_h, p_h)$ be the solution of coupled Stokes/Darcy model, and $(\underline{u}^H, p^h)$ be defined by and (27) on the fine grid. The following error estimates hold:*

$$\|\phi_h - \phi^h\|_{H_p} \lesssim H^2, \tag{27}$$

$$\|\boldsymbol{u}_h - \boldsymbol{u}^h\|_{H_f} \lesssim H^{3/2}, \tag{28}$$

$$\|p_h - p^h\|_Q \lesssim H^{3/2}. \tag{29}$$

Based on this algorithm, some other improvements have been made. For example, in [18–20], by sequentially solving the Stokes submodel and the Darcy submodel, the authors can make $\|\phi_h - \phi^h\|_{H_p}$, $\|\boldsymbol{u}_h - \boldsymbol{u}^h\|_{H_f}$, and $\|p_h - p^h\|_Q$ are all of order H^2. Furthermore, Hou constructed a new auxiliary problem [16] for the Darcy submodel, and proved that Mu and Xu's two-grid algorithm can retain $\|\boldsymbol{u}_h - \boldsymbol{u}^h\|_{H_f}$ and $\|p_h - p^h\|_Q$ order of H^2. It is remarkable that Mu and Xu's two-grid algorithm is naturally parallel and of optimal order, if h is of order H^2.

The extension to a multilevel decoupled algorithm can be found in [26]. The **Multilevel Algorithm** is as follows:

1. Solve the linear part of problem $(2.7)_H$ on a coarse grid: find $\underline{u}_H = (\boldsymbol{u}_H, \phi_H) \in \underline{X}_H \subset \underline{X}_h, p_H \in Q_H \subset Q_h$ such that

$$\begin{cases} a(\underline{u}_H, \underline{v}_H) + b(\underline{v}_H, p_H) = f(\underline{v}_H), \forall \underline{v}_H = (\boldsymbol{v}_H, \psi_H) \in \underline{X}_H, \\ b(\underline{u}_H, q_H) = 0, \forall q_H \in Q_H; \end{cases} \tag{30}$$

2. Set $h_0 = H$, for j = 1 to L,

   find $u^{h_j} = (\boldsymbol{u}^{h_j}, \phi^{h_j}) \in \underline{X}_{h_j}, p^{h_j} \in Q_{h_j}$ such that

$$\begin{cases} a\left(u^{h_j}, v_{h_j}\right) + b\left(v_{h_j}, p^{h_j}\right) = f\left(v_{h_j}\right) - a_\Gamma\left(u^{h_{j-1}}, v_{h_j}\right), \quad \forall v_{h_j} \in \underline{X}_{h_j}, \\ b\left(u^{h_j}, q_{h_j}\right) = 0, \quad \forall q_{h_j} \in Q_{h_j}. \end{cases} \tag{31}$$

   end.

In our multilevel algorithm, we refine the grid step by step, the coupled problem is only solved on the coarsest mesh, and linear decoupled subproblems are solved in parallel on successively refined meshes. We see that the algorithm is very effective and efficient. Moreover, the theory of the two grid algorithm guarantees that the approximation properties are good. As an illustration of the effectiveness of the multilevel algorithm [41], we present numerical results in **Table 2**.

In the following, we use steady state NS/Darcy model to illustrate how to apply two-level and multilevel methods to decouple the coupled nonlinear PDE models. The algorithm combines the two-level algorithms and the Newton-type linearization [4, 36, 42]. Our **Newton Type Linearization Based Two-level Algorithm** consists of the following three steps [17, 43].

1. Solve the coupled problem (19) on a coarse grid triangulation with the meshsize H: Find $\underline{u}_H = (\boldsymbol{u}_H, \phi_H) \in \underline{X}_H$ and $p_H \in Q_H$ such that

$$\begin{cases} a(\underline{u}_H, \underline{v}_H) + c(\boldsymbol{u}_H, \boldsymbol{u}_H, \boldsymbol{v}_H) + b(\boldsymbol{v}_H, p_H) = f(\underline{v}_H), \quad \forall \underline{v}_H = (\boldsymbol{v}_H, \psi_H) \in \underline{X}_H, \\ b(\boldsymbol{u}_H, q_H) = 0, \quad \forall q_H \in Q_H. \end{cases} \tag{32}$$

2. On a fine grid triangulation with the meshsize $h \leq H$, sequentially solve two decoupled and linearized local subproblems:

h	$\lvert\phi^h - \phi\rvert_1$	$\lvert\boldsymbol{u}^h - \boldsymbol{u}\rvert_1$	$\lvert v^h - v\rvert_1$	$\lVert p^h - p\rVert_0$
2^{-1}	4.592×10^{-2}	1.550×10^{-1}	1.066×10^{-1}	8.410×10^{-2}
2^{-2}	1.152×10^{-2}	3.958×10^{-2}	2.664×10^{-2}	1.752×10^{-2}
2^{-4}	7.280×10^{-4}	2.466×10^{-3}	1.652×10^{-3}	1.040×10^{-3}
2^{-8}	5.296×10^{-6}	9.981×10^{-6}	7.922×10^{-6}	1.694×10^{-5}

Table 2.
Errors between the solutions of multilevel algorithm and the exact solutions (second order discretization).

Step a. Solve a discrete Darcy problem in Ω_p: Find $\phi_h^* \in X_{p,h}$ such that

$$a_p(\phi_h^*, \psi_h) = \rho g\left(f_p, \psi_h\right)_{\Omega_p} + \rho g(\boldsymbol{u}_H \cdot \boldsymbol{n}_f, \psi_h)_\Gamma \quad \forall \psi_h \in X_{p,h}. \tag{33}$$

Step b. Solve a modified Navier–Stokes model using the Newton type linearization: Find $\boldsymbol{u}_h^* \in \boldsymbol{X}_{f,h}$ and $p_h^* \in Q_h$ such that

$$\begin{cases} a_f(\boldsymbol{u}_h^*, \boldsymbol{v}_h) + c(\boldsymbol{u}_H, \boldsymbol{u}_h^*, \boldsymbol{v}_h) + c(\boldsymbol{u}_h^*, \boldsymbol{u}_H, \boldsymbol{v}_h) + b(\boldsymbol{v}_h, p_h^*) = \left(\boldsymbol{f}_f, \boldsymbol{v}_h\right)_{\Omega_f} \\ -\rho g(\phi_h^*, \boldsymbol{v}_h \cdot \boldsymbol{n}_f)_\Gamma + c(\boldsymbol{u}_H, \boldsymbol{u}_H, \boldsymbol{v}_h) \quad \forall \boldsymbol{v}_h \in \boldsymbol{X}_{f,h}, \\ b(\boldsymbol{u}_h^*, q_h) = 0 \quad \forall q_h \in Q_h. \end{cases} \tag{34}$$

3. On the same fine grid triangulation, solve two subproblems by using the newly obtained solution.

Step a. Solve a discrete Darcy problem in Ω_p: Find $\phi^h \in X_{p,h}$ such that

$$a_p(\phi^h, \psi_h) = \rho g\left(f_p, \psi_h\right)_{\Omega_p} + \rho g(\boldsymbol{u}_h^* \cdot \boldsymbol{n}_f, \psi_h)_\Gamma \quad \forall \psi_h \in X_{p,h}. \tag{35}$$

Step b. Correct the solution of the fluid flow model: Find $\boldsymbol{u}^h \in \boldsymbol{X}_{f,h}$ and $p^h \in Q_h$ such that

$$\begin{cases} a_f(\boldsymbol{u}^h, \boldsymbol{v}_h) + c(\boldsymbol{u}_H, \boldsymbol{u}^h, \boldsymbol{v}_h) + c(\boldsymbol{u}^h, \boldsymbol{u}_H, \boldsymbol{v}_h) + b(\boldsymbol{v}_h, p^h) = \left(\boldsymbol{f}_f, \boldsymbol{v}_h\right)_{\Omega_f} \\ -\rho g(\phi^h, \boldsymbol{v}_h \cdot \boldsymbol{n}_f)_\Gamma + c(\boldsymbol{u}_H, \boldsymbol{u}_h^*, \boldsymbol{v}_h) + c(\boldsymbol{u}_h^*, \boldsymbol{u}_H - \boldsymbol{u}_h^*, \boldsymbol{v}_h) \quad \forall \boldsymbol{v}_h \in \boldsymbol{X}_{f,h}, \\ b(\boldsymbol{u}^h, q_h) = 0 \quad \forall q_h \in Q_h. \end{cases} \tag{36}$$

We remark here that the problem (36) and the problem (34) differ only in the right hand side. Similarly, the problem (35) and the problem (33) have the same stiffness matrix. In sum, the advantages of our work exist in that the scaling between the two meshsizes is better, the algorithm is decoupled and linear on the fine grid level and the two submodels in the last two steps share the same stiffness matrices.

For the coupled problem (19), by using the properties (16)–(18), the error estimates in the energy norm can be derived by using a fixed-point framework [4, 31]. Moreover, the Aubin-Nitsche duality argument can result in the L^2 error analysis of the problem (19). In summary, we have.

Lemma 1. *Let* $(\boldsymbol{u}, \phi, p) \in \boldsymbol{H}^{k+1}(\Omega_f) \times H^{k+1}(\Omega_p) \times H^k(\Omega_f)$ *be the solution of the Navier–Stokes/Darcy model (13) and* $(\boldsymbol{u}_h, \phi_h, p_h)$ *be the FE solution of (19). We assume that* ν *is sufficiently large and* h *is sufficiently small. There holds the following energy norm estimate for the problem (19).*

$$|\boldsymbol{u} - \boldsymbol{u}_h|_{1,\Omega_f} + |\phi - \phi_h|_{1,\Omega_p} + \|p - p_h\|_{0,\Omega_f} \lesssim h^k. \tag{37}$$

Moreover, we have the following L^2 error estimate:

$$\|\boldsymbol{u} - \boldsymbol{u}_h\|_{0,\Omega_f} + \|\phi - \phi_h\|_{0,\Omega_p} \lesssim h^{k+1}. \tag{38}$$

The energy norm estimate (37) is the Lemma 2 in [4]. The L^2 error estimate (38) corresponds to the Lemma 3 in [4]. Detailed proofs of these results can be found in [4].

The following lemma concludes the error estimate of $(\phi_h^*, \boldsymbol{u}_h^*, p_h^*)$ in the energy norm.

Lemma 2. (**Error analysis of the intermediate step two-level solution**) *Let $(\phi, \boldsymbol{u}, p)$ and $(\phi_h^*, \boldsymbol{u}_h^*, p_h^*)$ be defined by the problems (13) and (33)–(34), respectively. Under the assumptions of Lemma 1, there holds*

$$|\phi - \phi_h^*|_{1,\Omega_p} \lesssim H^{k+1} + h^k, \tag{39}$$

$$|\boldsymbol{u} - \boldsymbol{u}_h^*|_{1,\Omega_f} + \|p - p_h^*\|_{0,\Omega_f} \lesssim H^{k+1} + h^k. \tag{40}$$

$$\|\boldsymbol{u} - \boldsymbol{u}_h^*\|_{0,\Omega_f} \lesssim H^{2k+1} + H^{k+1}h + h^{k+1}. \tag{41}$$

From Lemma 1, we note that the optimal finite element solution error in the energy norm is of order $O\left(h^k\right)$. Combining the conclusions of this Lemma, we see that the intermediate step two-level solution error is still optimal in the energy norm if the scaling between the two meshsizes is taken to be $h = H^{\frac{k+1}{k}}$. The L^2 error analysis is extended from [4, 20, 31, 37]. Let **u** and ϕ be the nonsingular solution of (13). From Lemma 1, the optimal L^2 error for the finite element solution is of order $O\left(h^{k+1}\right)$. To make sure $\|\boldsymbol{u} - \boldsymbol{u}_h^*\|_{0,\Omega_f}$ is also of order $O\left(h^{k+1}\right)$, the scaling between the two grids has to be taken as $h = \max\left\{H^{\frac{k+1}{k}}, H^{\frac{2k+1}{k+1}}\right\}$. For instance, if $k = 1$, we have to set $h = H^{\frac{3}{2}}$ to make sure the L^2 error of $\boldsymbol{u}_h^*$ is optimal. We now show that the final step two-level solution is indeed a good approximation to the solution of problem (13) .

Theorem 2. (**Error analysis of the final step two-level solution**) *Let $(\phi, \boldsymbol{u}, p)$ and $(\phi^h, \boldsymbol{u}^h, p^h)$ be the solutions of (13) and (35)–(36) respectively. Under the assumptions of Lemma 1, the following error estimates hold:*

$$|\phi - \phi^h|_{1,\Omega_p} + |\boldsymbol{u} - \boldsymbol{u}^h|_{1,\Omega_f} + \|p - p^h\|_{0,\Omega_f} \lesssim H^{2k+1} + H^{k+1}h + h^k, \tag{42}$$

Proposition 1. *Let $(\phi, \boldsymbol{u}, p)$ and $(\phi^h, \boldsymbol{u}^h, p^h)$ be the solutions of (13) and (35)–(36) respectively. If we take $h = H^{\frac{2k+1}{k}}$ for $k = 1, 2$ and $h = H^{\frac{k+1}{k-1}}$ for $k \geq 3$, then there holds the following error estimate.*

$$|\phi - \phi^h|_{1,\Omega_p} + |\boldsymbol{u} - \boldsymbol{u}^h|_{1,\Omega_f} + \|p - p^h\|_{0,\Omega_f} \lesssim h^k. \tag{43}$$

Finally, we would like to make some comments on the mixed Stokes/Darcy model. We note that by dropping those trilinear terms (32), (34) and (36), our two-level algorithm can be naturally applied to the coupled Stokes/Darcy model. We note that the above algorithms can be naturally extended to multi-level algorithms by recursively calling the above two-level algorithms [17, 43]. The extension and the corresponding analysis can be found in [26, 43].

2.4 Decoupled algorithms by partitioned time schemes

The fully evolutionary Stokes/Darcy equations will be used as the model problem in this subsection to illustrate the partitioned time schemes. We neglect the fluid density and the porosity effects in this subsection. We review some decoupled methods that converge within a reasonable amount of time, and are stable when the physical parameters are small. More precisely, partitioned time methods can efficiently solve the surface subproblem and the subsurface subproblem separately.

For the estimate of the stability, we assume that the solution of the Stokes/Darcy problem is long-time regular [44]:

$$\begin{aligned} &\boldsymbol{u} \in \boldsymbol{W}^{2,\infty}(0,\infty; \boldsymbol{L}^2(\Omega_f)) \cap \boldsymbol{W}^{1,\infty}(0,\infty; \boldsymbol{H}^2(\Omega_f)), \\ &\phi \in W^{2,\infty}(0,\infty; L^2(\Omega_p)) \cap W^{1,\infty}(0,\infty; H^2(\Omega_p)), \\ &p \in L^{2,\infty}(0,\infty; H^1(\Omega_f)). \end{aligned} \tag{44}$$

The simplest time scheme for the evolutionary coupled Stokes/Darcy model is the **Backward Euler Algorithm**, which reads as: Given $(\boldsymbol{u}_h^n, p_h^n, \phi_h^n) \in \boldsymbol{X}_{f,h} \times Q_h \times X_{p,h}$, find $(\boldsymbol{u}_h^{n+1}, p_h^{n+1}, \phi_h^{n+1}) \in \boldsymbol{X}_{f,h} \times Q_h \times X_{p,h}$, such that for all $\boldsymbol{v}_h \in \boldsymbol{X}_{f,h}, q_h \in Q_h, \psi_h \in X_{p,h}$,

$$\begin{cases} \left(\dfrac{\boldsymbol{u}_h^{n+1} - \boldsymbol{u}_h^n}{\Delta t}, \boldsymbol{v}_h\right) + a_f(\boldsymbol{u}_h^{n+1}, \boldsymbol{v}_h) - (p_h^{n+1}, \nabla \cdot \boldsymbol{v}_h) + g(\boldsymbol{v}_h \cdot \boldsymbol{n}_f, \phi_h^{n+1})_\Gamma = \left(\boldsymbol{f}_f^{n+1}, \psi_h\right), \\ (q_h, \nabla \cdot \boldsymbol{u}_h^{n+1}) = 0, \end{cases} \tag{45}$$

$$gS_0\left(\frac{\phi_h^{n+1} - \phi_h^{n-1}}{\Delta t}, \boldsymbol{v}_h\right) + a_p(\phi_h^{n+1}, \psi_h) - g(\boldsymbol{u}_h^{n+1} \cdot \boldsymbol{n}_f, \psi_h)_\Gamma = g\left(f_p^{n+1}, \psi_h\right). \tag{46}$$

However, this scheme is fully coupled and each time step one has to solve a coupled system including both (45) and (46), although, on the other hand, this scheme enjoys the desirable strong stability and convergence properties. In [12], Mu and Zhu propose the following backward Euler forward Euler scheme and combine it with the two-level spatial discretization. We neglect the two-level spatial discretization in this presentation. Here, the Forward Euler means it discretizes the coupling term explicitly. **Backward Euler Forward Euler Scheme (BEFE)**: given $(\boldsymbol{u}_h^n, p_h^n, \phi_h^n) \in \boldsymbol{X}_{f,h} \times Q_h \times X_{p,h}$, find $(\boldsymbol{u}_h^{n+1}, p_h^{n+1}, \phi_h^{n+1}) \in \boldsymbol{X}_{f,h} \times Q_h \times X_{p,h}$, such that for all $\boldsymbol{v}_h \in \boldsymbol{X}_{f,h}, q_h \in Q_h, \psi_h \in X_{p,h}$,

$$\begin{cases} \left(\dfrac{\boldsymbol{u}_h^{n+1} - \boldsymbol{u}_h^n}{\Delta t}, \boldsymbol{v}_h\right) + a_f(\boldsymbol{u}_h^{n+1}, \boldsymbol{v}_h) - (p_h^{n+1}, \nabla \cdot \boldsymbol{v}_h) + g(\boldsymbol{v}_h \cdot \boldsymbol{n}_f, \phi_h^{n})_\Gamma = \left(\boldsymbol{f}_f^{n+1}, \psi_h\right), \\ (q_h, \nabla \cdot \boldsymbol{u}_h^{n+1}) = 0, \end{cases} \tag{47}$$

$$gS_0\left(\frac{\phi_h^{n+1} - \phi_h^{n}}{\Delta t}, \boldsymbol{v}_h\right) + a_p(\phi_h^{n+1}, \psi_h) - g(\boldsymbol{u}_h^{n} \cdot \boldsymbol{n}_f, \psi_h)_\Gamma = g\left(f_p^{n+1}, \psi_h\right). \tag{48}$$

The analysis of this can be found in [12, 45]. In particular, the longtime stability (cf. (51)) of BEFE method was proved in [45] in the sense that no form of Gronwall's inequality was used.

Theorem 3. *Assume the following time step condition is satisfied*

$$\Delta t \lesssim min\left\{\nu k_{min}^2, S_0 \nu^2 k_{min}\right\}, \tag{49}$$

Then, BELF algorithm achieves the optimal convergence rate uniformly in time. The solution of the BEFE method satisfies the uniform in time error estimates:

i. If $\boldsymbol{f}_f \in L^\infty(0,+\infty;\boldsymbol{L}^2(\Omega_f)), f_p \in L^\infty(0,+\infty;L^2(\Omega_p))$, then

$$\|\boldsymbol{u}_h^n\|^2 + \|\phi_h^n\|^2 \le C, \quad \forall n \ge 0. \tag{50}$$

ii. If $\|\boldsymbol{f}_f\|_{L^\infty(0,+\infty;\boldsymbol{L}^2(\Omega_f))}, \|f_p\|_{L^\infty(0,+\infty;L^2(\Omega_p))}$ are uniformly bounded in Δt, then

$$\|\boldsymbol{u}_h^n\|^2 + \|\phi_h^n\|^2 + \Delta t\sum_{l=0}^{n}\left(\|\nabla \boldsymbol{u}_h^l\|^2 + \|\nabla\phi_h^l\|^2\right) \le C, \quad \forall n \ge 0. \tag{51}$$

The advantage of this scheme is that it is parallel. As revealed in the time step restriction (50), the disadvantage of this method is that it may become highly unstable when the parameters S_0 and k_{min} are small. Another way for uncoupling surface/subsurface flow models is using splitting schemes which require sequential subproblem solves at each time step [46]. As an example, we note that in solving (49), one can replace $\boldsymbol{u}_h^n$ by using the most updated solution obtained in the Stokes step. This will lead to Backward Euler time-split scheme [46]. We skip the details of this time-split method, interested readers can refer to [46]. By this way, one can design different sequential splitting schemes. Noting that the BEFE method is only of first order, in some other decoupled Implicit-explicit (IMEX) methods, one can combine of the three level implicit method with the coupling terms treated by the explicit method to achieve high order. For example, Crank–Nicolson Leap-Frog method [47, 48], second-order backward-differentiation with Gear's extrapolation, Adam-Moulton-Bashforth [49]. We present one of them: the **Crank–Nicolson Leap-Frog Method** for the evolutionary Stokes/Darcy model: given $(\boldsymbol{u}_h^{n-1}, p_h^{n-1}, \phi_h^{n-1}), (\boldsymbol{u}_h^n, P_h^n, \phi_h^n) \in \boldsymbol{X}_{f,h} \times Q_h \times X_{p,h}$, find $(\boldsymbol{u}_h^{n+1}, p_h^{n+1}, \phi_h^{n+1}) \in \boldsymbol{X}_{f,h} \times Q_h \times X_{p,h}$, such that for all $\boldsymbol{v}_h \in \boldsymbol{X}_{f,h}, q_h \in Q_h, \psi_h \in X_{p,h}$,

$$\begin{cases} \left(\dfrac{\boldsymbol{u}_h^{n+1} - \boldsymbol{u}_h^{n-1}}{2\Delta t}, \boldsymbol{v}_h\right) + \left(\nabla\cdot\left(\dfrac{\boldsymbol{u}_h^{n+1} - \boldsymbol{u}_h^{n-1}}{2\Delta t}, \boldsymbol{v}_h\right), \nabla\cdot\boldsymbol{v}_h\right) + a_f\left(\dfrac{\boldsymbol{u}_h^{n+1} + \boldsymbol{u}_h^{n-1}}{2}, \boldsymbol{v}_h\right) \\ -\left(\dfrac{p_h^{n+1} - p_h^{n-1}}{2}, \nabla\cdot\boldsymbol{v}_h\right) + g(\boldsymbol{v}_h, \phi_h^n)_\Gamma = \left(\boldsymbol{f}_f^n, \boldsymbol{v}_h\right), \\ \left(q_h, \nabla\cdot\dfrac{\boldsymbol{u}_h^{n+1} - \boldsymbol{u}_h^{n-1}}{2}\right) = 0, \end{cases} \tag{52}$$

$$gS_0\left(\frac{\phi_h^{n+1} - \phi_h^{n-1}}{2\Delta t}, \phi_h\right) + a_p\left(\frac{\phi_h^{n+1} + \phi_h^{n-1}}{2}, \psi_h\right)_h - g(\boldsymbol{u}_h^n, \psi_h)_\Gamma = g\left(f_p^n, \psi_h\right). \tag{53}$$

The Crank–Nicolson-Leap-Frog possesses strong stability and convergence properties [47, 48]. Most importantly, the time-step condition for the scheme does not depend on κ_{min}.

For the numerical experiments of these partitioned time schemes, we refer the readers to [12].

3. FSI models and decoupled algorithms

FSI models include a fluid model whose general form is (1), a structure model, plus certain interface conditions that describe the interaction mechanism (see **Figure 2** for an illustration of FSI models in the reference configuration and the deformed configuration). To differentiate the notations in different subdomains, we will use a subscript "*f*" to denote the variables in the fluid domain, and a subscript "*s*" to denote the variables in the structure domain.

In general, the structure model reads as:

$$\begin{cases} \rho_s\left(\dfrac{\partial \boldsymbol{u}_s}{\partial t} + (\boldsymbol{u}_s \cdot \nabla)\boldsymbol{u}_s\right) - \boldsymbol{div}\boldsymbol{T}(\boldsymbol{u}_s, p_s) = \boldsymbol{f}_s, \quad \forall \boldsymbol{x} \in \Omega_s, \\ \mathrm{div}\boldsymbol{u}_s = 0, \quad \forall \boldsymbol{x} \in \Omega_s. \end{cases} \tag{54}$$

Here, ρ_s is the density of the structure, $\boldsymbol{u}_s$ is the structure velocity, p_s is the structure pressure. In structure mechanics, the displacement $\boldsymbol{d}$ is usually used as a primary variable ($\dot{\boldsymbol{d}} = \boldsymbol{u}_s$), and the stress term in the linear case can be described by using Hooke's law. As the structure model is usually based on Lagrangian coordinates, researchers usually introduce the so-called Arbitrary Lagrangian Eulerian description for FSI models. In some special cases of FSI models, one can apply the simplified structure model such as 1D structure model or linear elasticity model for structure part, and the simplified fluid model such as linear Stokes or inviscid flow model for fluid part.

The fluid motion and the structure motion are coupled through certain interface conditions that describe the compatibility of the kinematics and transactions at the fluid–structure interface. For applications with non-slip interface conditions, both velocity and normal stress are continuous across the interface Γ, which may be described as

$$\begin{cases} \boldsymbol{u}_f = \boldsymbol{u}_s, \quad \text{on } \Gamma, \\ \boldsymbol{T}\left(\boldsymbol{u}_f, p_f\right)\boldsymbol{n} = \boldsymbol{T}(\boldsymbol{u}_s, p_s)\boldsymbol{n}, \text{ on } \Gamma. \end{cases} \tag{55}$$

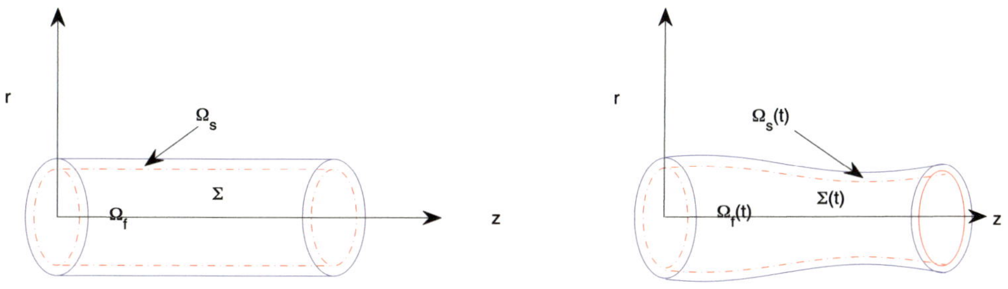

Figure 2.
An illustration of fluid structure interaction: the reference configuration (left) and the deformed configuration (right).

Here, $\boldsymbol{n}$ denotes one of $\boldsymbol{n}_f$ and $\boldsymbol{n}_s$. With suitable initial conditions and boundary conditions such as fluid flux boundary condition and structure Dirichlet boundary condition, the FSI models are complete.

There have been many advanced numerical methods for various FSI models. Our focus here is a linear fluid model coupled with a thin wall structure. The reason we choose this model is that this kind of models, if calculated using the standard (Dirichlet-Neumann) explicit decoupling schemes will lead to the so-called added mass effect [50]. Moreover, the algorithms dealing with the added-mass difficulties are the most exciting development in this direction. Therefore, we consider dropping the nonlinear term in (1), and consider the following 1D structure model:

$$\begin{cases} \rho_s \varepsilon \partial_t \dot{\boldsymbol{d}} + \boldsymbol{L}_s\left(\boldsymbol{d}, \dot{\boldsymbol{d}}\right) = -\boldsymbol{T}\left(\boldsymbol{u}_f, p_f\right)\boldsymbol{n}, & \text{on} \quad \Gamma, \\ \boldsymbol{u}_s = \dot{\boldsymbol{d}} \quad \text{on} \quad \Gamma. \end{cases} \tag{56}$$

Here, ε denotes the structure thickness, $\boldsymbol{L}_s\left(\boldsymbol{d}, \dot{\boldsymbol{d}}\right)$ denotes the operator in the structure model, which may include both the elastic term and the damping term [14, 15].

3.1 Partitioned algorithms for FSI models

First of all, we comment here that the decoupled preconditioning techniques can also be naturally applied to FSI models. In the preconditioning step, one can apply either the Multigrid approach [51] or domain decomposition methods [52].

In this presentation, we focus on the two most recent approaches for the linear Stokes model coupled with thin wall structure. The first approach is called partitioned Robin-Neumann scheme, in which the fluid subproblem is imposed with Robin boundary condition at the interface while the structure subproblem is imposed with Neumann boundary conditioner at the interface [14, 15]. The second approach is called kinetically coupled β-scheme, which is actually decoupled in the sense that computations are realized subdomain by subdomain [13, 21]. The derivation of the β-scheme is based on operator splitting.

Partitioned Robin-Neumann scheme:

1. Given the initial solution $\boldsymbol{u}_f^0$,p_f^0 and $\boldsymbol{d}^0$.

2. For $m = 0, 1, 2, 3, \ldots, N-1$,

    - **Fluid step**: find $\boldsymbol{u}_f^{m+1}$ and p_f^{m+1} such that

$$\begin{cases} \frac{\rho_f}{\Delta t}\left(\boldsymbol{u}_f^{m+1} - \boldsymbol{u}_f^m\right) - \operatorname{div}\boldsymbol{T}\left(\boldsymbol{u}_f^{m+1}, p_f^{m+1}\right) = \boldsymbol{0} & \text{in} \quad \Omega_f, \\ \operatorname{div}\boldsymbol{u}_f^{m+1} = 0 & \text{in} \quad \Omega_f, \\ \boldsymbol{T}\left(\boldsymbol{u}_f^{m+1}, p_f^{m+1}\right)\boldsymbol{n} + \frac{\rho_s \varepsilon}{\Delta t}\boldsymbol{u}_f^{m+1} = \frac{\rho_s \varepsilon}{\Delta t}\dot{\boldsymbol{d}}^m + \boldsymbol{T}\left(\boldsymbol{u}_f^m, p^m\right)\boldsymbol{n} & \text{on} \quad \Gamma. \end{cases} \tag{57}$$

    - **Structure step**: find $\mathbf{d}_s^{m+1}$ such that

$$\begin{cases} \frac{\rho_s \varepsilon}{\Delta t}\left(\dot{\boldsymbol{d}}^{m+1} - \dot{\boldsymbol{d}}^{m}\right) + \mathbf{L}_s\left(\boldsymbol{d}^{m+1}, \dot{\boldsymbol{d}}^{m+1}\right) = -\boldsymbol{T}\left(\boldsymbol{u}_f^{m+1}, p_f^{m+1}\right)\boldsymbol{n} & \text{on } \Gamma, \\ \dot{\boldsymbol{d}}^{m+1} = \frac{1}{\Delta t}\left(\boldsymbol{d}^{m+1} - \boldsymbol{d}^{m}\right) & \text{on } \Gamma,. \end{cases} \tag{58}$$

3. End

Here, $\frac{\rho_s \varepsilon}{\Delta t}$ is treated as a Robin coefficient. These partitioned iterative methods were firstly introduced in [53], as added-mass free alternatives to the standard Dirichlet-Neumann scheme. Some extensions and generalizations can be found in [14, 15].

Different from the partitioned Robin-Neumann scheme. The kinematically coupled β-**scheme** for the time-discrete problem is given as follows. The stability and the convergence rate of this scheme are analyzed in [13, 21].

1. Given the initial solution $\boldsymbol{u}_f^0, p_f^0$ and $\boldsymbol{d}^0$.

2. For $m = 0, 1, 2, 3, \ldots, N-1$,

   - **Structure step:** find $\tilde{\boldsymbol{u}}_s^{m+1}$ such that

$$\begin{cases} \rho_s \varepsilon \frac{\tilde{\boldsymbol{u}}_s^{m+1} - \boldsymbol{u}_s^{m}}{\Delta t} + \mathbf{L}_s\left(\boldsymbol{d}^{m+1}, \dot{\boldsymbol{d}}^{m+1}\right) = -\beta \sigma_f\left(\boldsymbol{u}_f^{m}, p_f^{m}\right)\boldsymbol{n} & \text{on } \Gamma, \\ \dot{\boldsymbol{d}}^{m+1} = \tilde{\boldsymbol{u}}_s^{m+1}, \boldsymbol{d}^{m+1} = \boldsymbol{d}^{m} + \Delta t \tilde{\boldsymbol{u}}_s^{m+1} & \text{on } \Gamma. \end{cases} \tag{59}$$

   - **Fluid step:** find $\boldsymbol{u}_f^{m+1}$, p_f^{m+1} and $\boldsymbol{u}_s^{m+1}$ such that

$$\begin{cases} \frac{\rho_f}{\Delta t}\left(\boldsymbol{u}_f^{m+1} - \boldsymbol{u}_f^{m}\right) - \operatorname{div} \sigma_f\left(\boldsymbol{u}_f^{m+1}, p_f^{m+1}\right) = \mathbf{0} & \text{in } \Omega_f, \\ \operatorname{div} \boldsymbol{u}_f^{m+1} = 0 & \text{in } \Omega_f, \\ \rho_s \varepsilon \frac{\boldsymbol{u}_s^{m+1} - \tilde{\boldsymbol{u}}_s^{m+1}}{\Delta t} = -\sigma_f\left(\boldsymbol{u}_f^{m+1}, p_f^{m+1}\right)\boldsymbol{n} + \beta \sigma_f\left(\boldsymbol{u}_f^{m}, p_f^{m}\right)\boldsymbol{n} & \text{on } \Gamma, \\ \boldsymbol{u}_f^{m+1} = \boldsymbol{u}_s^{m+1} & \text{on } \Gamma. \end{cases} \tag{60}$$

3. End.

3.2 Multirate time step approach for FSI models

Due to different time scales in many FSI problems, it is natural and essential to develop multirate time-stepping schemes [22, 23] that mimic the physical phenomena. For illustration, we will examine the application of the multirate technique to the β-scheme, since similar performance is observed for both the Robin-Neumann scheme and the β-scheme in numerical experiments. Furthermore, the decoupled multirate β-scheme can be extended to more general FSI problems involving nonlinearity, irregular domains, and large structural deformations.

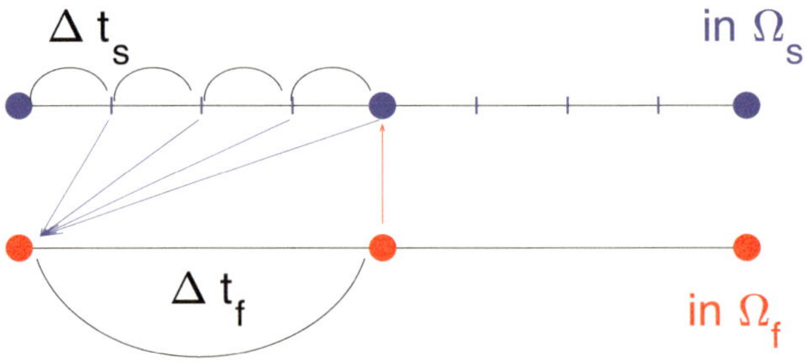

Figure 3.
An illustration of a multirate time stepping technique.

In order to apply multirate time-stepping scheme to FSI problems, a question is in which region the larger time step size should be used. Numerical experiments suggest that the version with a larger time step size in the fluid solver (cf. **Figure 3**) results in a better accuracy. The corresponding method is described in the following. We comment here that the multirate β-scheme is nothing else but the β-scheme itself when the time-step ratio $r = 1$.

Multirate time-stepping β-scheme:

1. Given the initial solution $\boldsymbol{u}_f^0$,p_f^0 and $\boldsymbol{d}^0$.

2. For $k = 0, 1, 2, 3, \dots, N-1,$ set $m_k = r \cdot k$.

  - **Structure steps:** for $m = m_k, m_k + 1, m_k + 2, \dots, m_{k+1} - 1,$

$$\begin{cases} \rho_s \varepsilon \dfrac{\tilde{\boldsymbol{u}}_s^{m+1} - \boldsymbol{u}_s^m}{\Delta t_s} + \boldsymbol{L}_s\left(\boldsymbol{d}^{m+1}, \dot{\boldsymbol{d}}^{m+1}\right) = -\beta \sigma_f\left(\boldsymbol{u}_f^{m_k}, p_f^{m_k}\right)\boldsymbol{n} & \text{on } \Gamma, \\ \dot{\boldsymbol{d}}^{m+1} = \tilde{\boldsymbol{u}}_s^{m+1}, \boldsymbol{d}^{m+1} = \boldsymbol{d}^m + \Delta t_s \tilde{\boldsymbol{u}}_s^{m+1} & \text{on } \Gamma. \end{cases} \tag{61}$$

  - **Fluid step:**

$$\begin{cases} \dfrac{\rho_f}{\Delta t_f}\left(\boldsymbol{u}_f^{m_{k+1}} - \boldsymbol{u}_f^{m_k}\right) - \operatorname{div} \sigma_f\left(\boldsymbol{u}_f^{m_{k+1}}, p_f^{m_{k+1}}\right) = \boldsymbol{0} & \text{in } \Omega_f, \\ \operatorname{div} \boldsymbol{u}_f^{m_{k+1}} = 0 & \text{in } \Omega_f, \\ \rho_s \varepsilon \dfrac{\boldsymbol{u}_s^{m_{k+1}} - \tilde{\boldsymbol{u}}_s^{m_{k+1}}}{\Delta t_f} = -\sigma_f\left(\boldsymbol{u}_f^{m_{k+1}}, p_f^{m_{k+1}}\right)\boldsymbol{n} + \beta \sigma_f\left(\boldsymbol{u}_f^{m_k}, p_f^{m_k}\right)\boldsymbol{n} & \text{on } \Gamma, \\ \boldsymbol{u}_f^{m_{k+1}} = \boldsymbol{u}_s^{m_{k+1}} & \text{on } \Gamma. \end{cases} \tag{62}$$

3. End.

3.3 Numerical experiment

In this subsection, we present numerical experiments to demonstrate the convergence and stability performance of the multirate β-scheme (60)–(61) for coupling a Stokes flow with a thin-walled structure by using a benchmark model. The model consists of a *2-D* rectangular fluid domain $\Omega_f = [0, L] \times [0, R]$ with $L = 6$ and $R = 0.5$ and a *1-D* structure domain $\Gamma = [0, L] \times R$ that meanwhile also plays the role of the fluid–solid interface, as shown in **Figure 2**. The displacement of the interface is assumed to be infinitesimal and the Reynolds number in the fluid is assumed to be small (**Figure 4**). All the quantities will be given in terms of the centimeter-gram-second (CGS) system of units.

The physical parameters are set as follows: $\rho_f = 1.0$, $\rho_s = 1.1$, $\mu = 0.035$; $\mathbf{L}_s\left(\mathbf{d}, \dot{\mathbf{d}}\right) = c_1 \partial_x^2 \mathbf{d} + c_0 \mathbf{d}$, where $c_1 = \frac{E\varepsilon}{2(1+\nu)}$, $c_0 = \frac{E\varepsilon}{R^2(1-\nu^2)}$ with $\varepsilon = 0.1$, the Poisson ratio $\nu = 0.5$ and the Young's modulus $E = 0.75 \cdot 10^6$. A pressure-wave

$$P(t) = P_{max}(1 - \cos(2t\pi/T^*))/2 \quad \text{with} \quad P_{max} = 2 \cdot 10^4, \tag{63}$$

is prescribed on the fluid inlet boundary for $T^* = 5 \cdot 10^{-3}$ (seconds). Zero traction is enforced on the fluid outlet boundary and no-slip condition is imposed on the lower boundary $y = 0$. For the solid, the two endpoints are fixed with $\mathbf{d} = \mathbf{0}$ at $x = 0$ and $x = 6$.

The first experiment is set up to compare the Robin-Neumann scheme with the β-scheme, the two stable decoupled methods recently developed for the benchmark model. **Figure 5** displays the displacements computed by the Robin-Neumann scheme and the β-scheme, together with the coupled implicit scheme for reference, where the

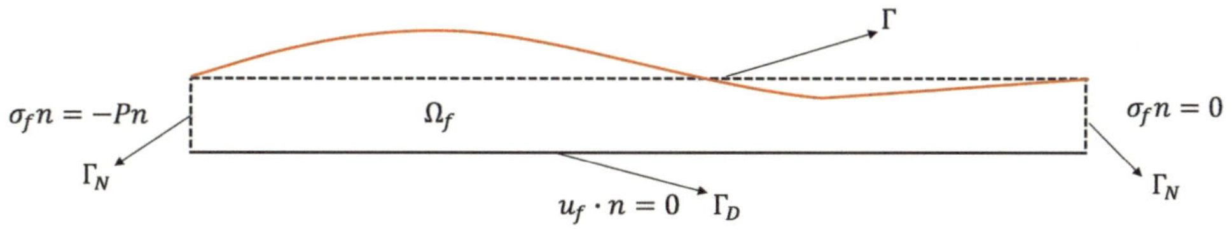

Figure 4.
Geometrical configuration.

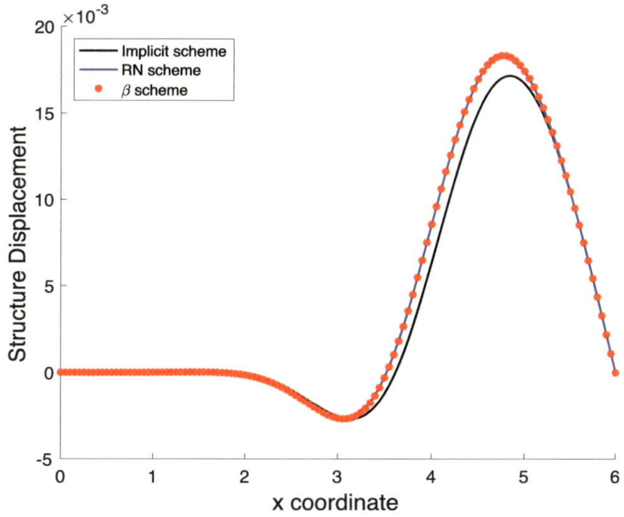

Figure 5.
Comparisons of the numerical results obtained by the coupled implicit scheme, the RN scheme, and the β-scheme under the setting: $h = 0.05$ and $\Delta t_s = 10^{-4}$.

mesh size and the time step size are $h = 0.05$ and $\Delta t = 10^{-4}$. It is clearly seen that both decoupled schemes converge as well as the coupled implicit scheme. Moreover, little difference is observed between the two decoupled schemes numerically. This suggests we focus on the β-scheme for investigating the multirate time-step technique.

We then conduct numerical experiments to investigate the effects of the time-step ratio r. **Figure 6** illustrates that a larger time step size in the fluid part results in a more accurate numerical solution than that obtained by using a larger step size in the structure part. In the test, we fix $h = 0.1$ and $\Delta t = 10^{-5}$. In addition, we observe that, when $\frac{\Delta t_s}{\Delta t_f}$ is further increased to be $\frac{\Delta t_s}{\Delta t_f} = 5$ or $\frac{\Delta t_s}{\Delta t_f} = 10$, there are substantial numerical instability. This screens out the possibility of using a larger time step size in the structure part.

To examine how the stability and approximation are affected when the time step in the fluid region is too large, we fix the time step Δt_s and h while varying the time-step ratio $r = 1, 5, 10, 20, 50$. **Figure 7** displays the computed displacements at $t = 0.015$ with the structure time step size $\Delta t_s = 10^{-5}$, the mesh size $h = 0.1$ (left) and $h = 0.01$ (right). In the left figure, we observe that the structure displacements computed by using $r = 1, 2, 5, 10$ approximate very well to that by using the coupled implicit scheme. To further investigate the stability and the convergence of the multirate β-scheme, in the right part of **Figure 7**, we reduce the mesh size to be $h = 0.01$ while fixing the time step size. The numerical results confirm that the multirate β-scheme is still stable even the time-step ratio is reasonably large.

In **Figure 8**, we present the numerical results of the fluid pressure distribution at $t = 0.005, 0.01, 0.015$. From the top to the bottom, numerical results are: a reference solution by the coupled implicit scheme, the numerical solution by the β-scheme, and the solution by the multirate β-scheme with $r = 10$. By comparing the results, we see that the multirate β-scheme provides a very good approximation.

In order to examine the order of convergence, we start with $h = 0.1$ and $\Delta t_s = 0.0001$, and then refine the mesh size by a factor of 2 and the time step size by a factor of 4. The space–time size settings are:

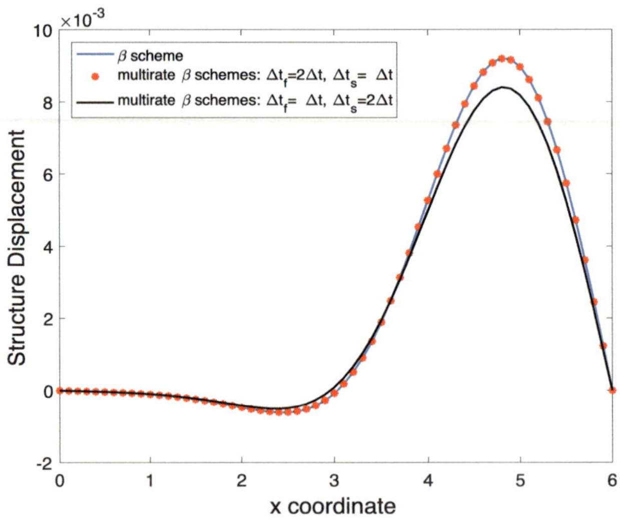

Figure 6.
Comparison of the β-scheme and the multirate β-schemes with two different time-step ratios ($h = 0.1$ and $\Delta t = 10^{-5}$ are fixed).

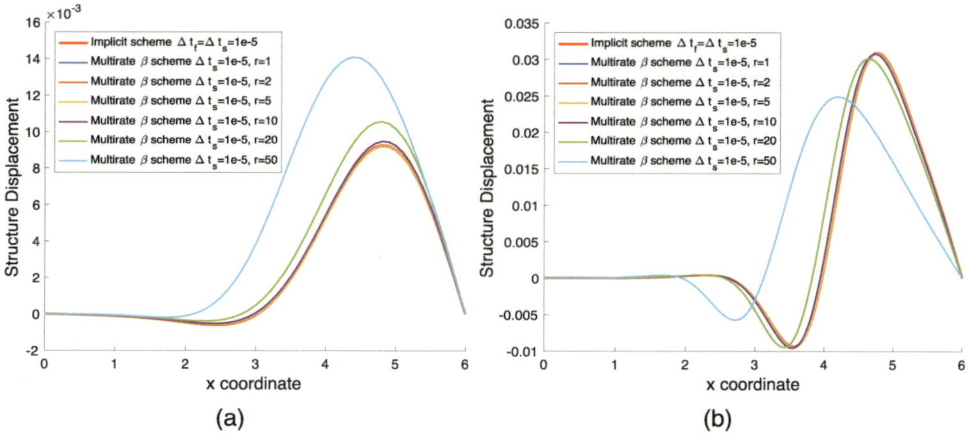

Figure 7.
Numerical displacements under the settings: $h = 0.1$ (left) $h = 0.01$ (right) and $\Delta t_s = 10^{-5}$.

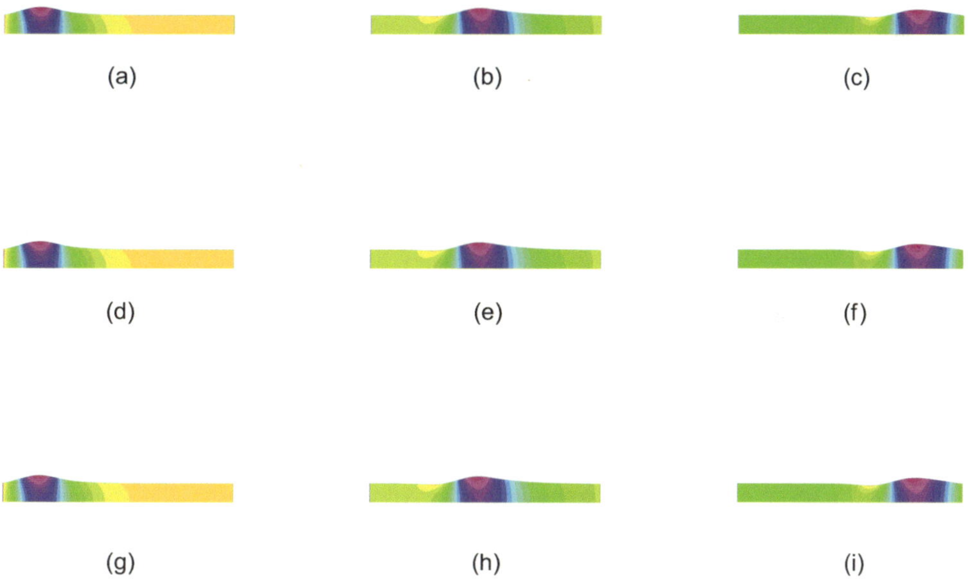

Figure 8.
Fluid pressure distribution at $t = 0.005, 0.010, 0.015$ obtained by the coupled implicit scheme (top), the multirate β-scheme with $r = 1$ (middle) and $r = 10$ (bottom) with $h = 0.01$ and $\Delta t_s = 0.00001$. (a) t = 0.005, (b) t = 0.010, (c) t = 0.015, (d) t = 0.005, (e) t = 0.010, (f) t = 0.015, (g) t = 0.005, (h) t = 0.010, (i) t = 0.015.

$$\{h, \Delta t_s\}^i = \left\{0.1 \cdot (0.5)^i, 0.0001 \cdot (0.25)^i\right\}, \quad i = 0, 1, 2, 3, 4. \tag{64}$$

We compare the numerical solutions of the multirate β-scheme with the reference solution. The reference solution is computed by using the coupled implicit scheme with a high space–time grid resolution $\left(h = 3.125 \times 10^{-3}, \Delta t = 10^{-6}\right)$ as that in [14]. In the multirate scheme, $r = 1$ and $r = 10$. The relative errors of the primary variables $(\boldsymbol{u}_f, p_f$ and $\boldsymbol{d})$ at $t = 0.015$ are displayed in **Figure 9**. From the comparisons, we see that the numerical errors are approximately reduced by a factor of 4 as the mesh size

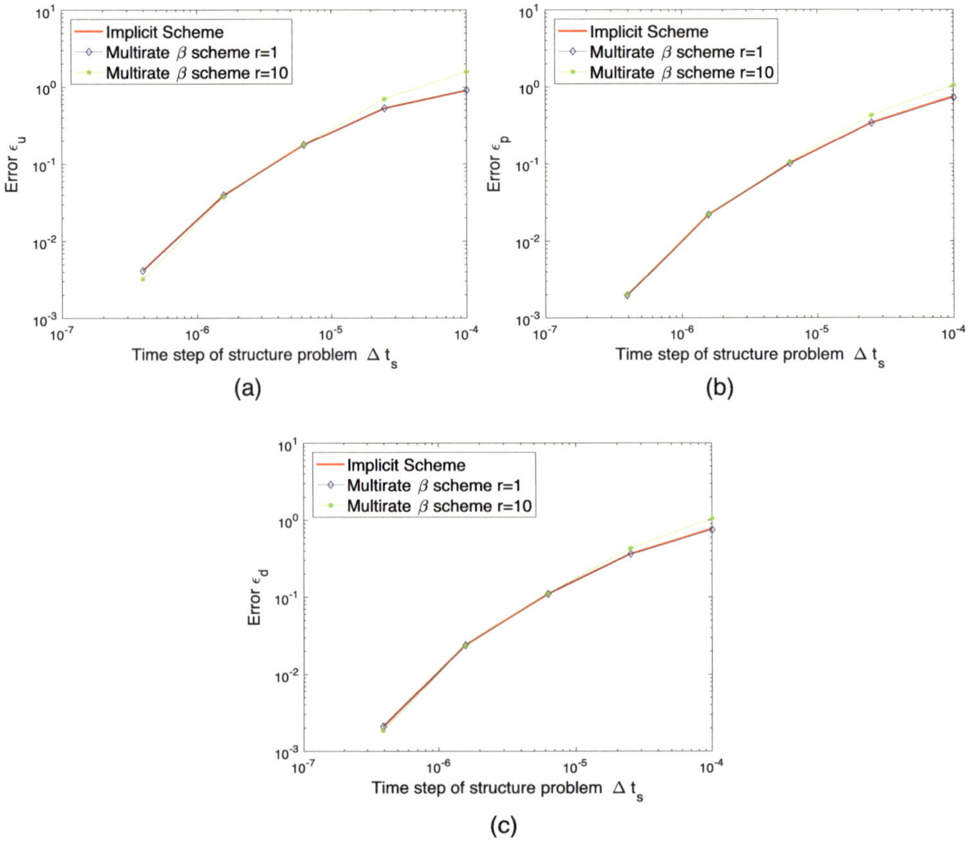

Figure 9.
Relative errors of primary variables with the spacing h and time step size Δt_s in (64). (a) Relative error of uf, (b) Relative error of pf, (c) Relative error of d.

and the time step size are refined once. Therefore, the multirate β-scheme is of a second order in h and a first order in t.

Finally, in order to demonstrate the advantage of the multirate β-scheme, we compare in **Table 3** the CPU times of the concerned numerical algorithms under various settings. We fixed $\Delta t_s = 10^{-5}$ and varying the mesh sizes as $h = \frac{1}{10}, \frac{1}{20}, \frac{1}{40}, \frac{1}{80}, \frac{1}{160}$. From the table, it is observed that the multirate β-scheme takes much less computational cost than that of the coupled implicit scheme, particularly when r is large.

	Implicit scheme	Multirate β-scheme $r = 1$	Multirate β-scheme $r = 10$
$h = \frac{1}{10}$	14.90	4.02	0.74
$h = \frac{1}{20}$	48.64	16.00	2.82
$h = \frac{1}{40}$	179.83	66.67	11.6
$h = \frac{1}{80}$	797.76	297.96	49.23
$h = \frac{1}{160}$	3165.26	1270.30	206.32

Table 3.
CPU times (in seconds) for the coupled implicit scheme and the multirate β-scheme (with $r = 1$ or 10) under different settings of mesh sizes ($\Delta t_s = 10^{-5}$ is fixed).

4. Concluding remarks

In numerical methods for coupled multidomain PDE models, there are usually two types of methods: the monolithic methods and the partitioned (or decoupled) methods. The monolithic methods usually require a code developed for the coupled problem being solved globally. In contrast, the partitioned approach preserves software modularity because one can use existing subdomain solvers. The advantages of using monolithic methods exist in that they provide better approximation accuracy and usually have better stability than the decoupled methods. The drawback is that the computation costs based on the monolithic approaches are usually high. In comparison, the partitioned approaches allow reusing existing software which is an attractive advantage. However, the accuracy and stability of the partitioned method need to be taken into consideration. Nevertheless, from our research works, we also note that one can apply decoupling techniques in monolithic methods, for example, decoupled preconditioners and schemes which are parallel in time. On the other hand, in partitioned algorithms, one can also try to apply coupling numerical techniques such as extrapolations using previous time-step solutions, interpolation using the coarse-grid solution, or extrapolations between subdomain solutions. In this work, we review the decoupling algorithms for the coupled models of fluid flow interacting with porous media flow and FSI models. We show how to decouple the coupled PDE models using preconditioning, two-level and multi-level algorithms, and partitioned time schemes. We attach importance to the decoupling numerical techniques while also emphasizing how to preserve the coupled multidomain physics features. This review provides a general framework for designing decoupled algorithms for coupled PDE models and exhibits the philosophy of the interplays between PDE models and the corresponding numerical methods.

Funding

Mingchao Cai's work was supported in part by the NSF Awards (170023, 1831950), NIH-BUILD grant through UL1GM118973, NIH-RCMI grant through U54MD013376, and XSEDE HPC resource TG-MCH200022. Mo Mu's work was supported in part by the HongKong RGC CERG HKUST16301218.

Author details

Mingchao Cai[1], Mo Mu[2*] and Lian Zhang[3]

1 Department of Mathematics, Morgan State University, Baltimore, MD, USA

2 Department of Mathematics, The Hong Kong University of Science and Technology, Hong Kong

3 In-Chao Institute Ltd, Shenzhen, China

*Address all correspondence to: mamu@ust.hk

References

[1] Beavers G, Joseph D. Boundary conditions at a naturally permeable wall. Journal of Fluid Mechanics. 1967;**30**(1): 197-207

[2] Saffman PG. On the boundary condition at the surface of a porous medium. Studies in Applied Mathematics. 1971;**50**(2):93-101

[3] Jäger W, Mikelić A. On the interface boundary condition of Beavers, Joseph, and Saffman. SIAM Journal on Applied Mathematics. 2000;**60**:1111-1127

[4] Cai M, Mu M, Xu J. Numerical solution to a mixed Navier-Stokes/Darcy model by the two-grid approach. SIAM Journal on Numerical Analysis. 2009;**47** (5):3325-3338

[5] Cesmelioglu A, Girault V, Rivière B. Time-dependent coupling of Navier-Stokes and Darcy flows. ESAIM: Mathematical Modelling and Numerical Analysis. 2013;**47**:539-554

[6] Girault V, Riviére B. DG approximation of coupled Navier-Stokes and Darcy equations by Beaver-Joseph-Saffman interface condition. SIAM Journal on Numerical Analysis. 2009;**47**: 2052-2089

[7] Layton WJ, Schieweck F, Yotov I. Coupling fluid flow with porous media flow. SIAM Journal on Numerical Analysis. 2003;**40**(6):2195-2218

[8] Layton W, Lenferink H. A multilevel mesh independence principle for the Navier-Stokes equations. SIAM Journal on Numerical Analysis. 1996;**33**(1):17-30

[9] Cai M. Modeling and Numerical Simulation for the Coupling of Surface Flow with Subsurface Flow, [PhD thesis]. Hong Kong University of Science and Technology; 2008

[10] Li Z. An augmented cartesian grid method for Stokes¨CDarcy fluid¨Cstructure interactions. International Journal for Numerical Methods in Engineering. 2016;**106**(7): 556-575

[11] Mu M, Xu J. A two-grid method of a mixed stokes-Darcy model for coupling fluid flow with porous media flow. SIAM Journal on Numerical Analysis. 2007;**45** (5):1801-1813

[12] Mu M, Zhu X. Decoupled schemes for a non-stationary mixed Stokes-Darcy model. Mathematics of Computation. 2010;**79**(270):707-731

[13] Bukac M, Muha B. Stability and convergence analysis of the extensions of the kinematically coupled scheme for the fluid-structure interaction. SIAM Journal on Numerical Analysis. 2016;**54**(5): 3032-3061

[14] Fernández M, Mullaert J, Vidrascu M. Explicit Robin–Neumann schemes for the coupling of incompressible fluids with thin-walled structures. Computer Methods in Applied Mechanics and Engineering. 2013;**267**:566-593

[15] Fernández M, Mullaert J, Vidrascu M. Generalized Robin–Neumann explicit coupling schemes for incompressible fluid-structure interaction: Stability analysis and numerics. International Journal for Numerical Methods in Engineering. 2015;**101**(3):199-229

[16] Hou Y. Optimal error estimates of a decoupled scheme based on two-grid finite element for mixed StokesCDarcy model. Applied Mathematics Letters. 2016;**57**:90-96

[17] Huang P, Cai M, Wang F. A Newton type linearization based two grid method

for coupling fluid flow with porous media flow. Applied Numerical Mathematics. 2016;**106**:182-198

[18] Zuo L, Hou Y. A decoupling two-grid algorithm for the mixed Stokes-Darcy model with the Beavers-Joseph interface condition. Numerical Methods for Partial Differential Equations. 2014; **30**(3):1066-1082

[19] Zuo L, Hou Y. Numerical analysis for the mixed Navier–Stokes and Darcy problem with the Beavers–Joseph interface condition. Numerical Methods for Partial Differential Equations. 2015; **31**(4):1009-1030

[20] Zhang T, Yuan J. Two novel decoupling algorithms for the steady Stokes-Darcy model based on two-grid discretizations. Discrete and Continuous Dynamical Systems—Series B. 2014;**19** (3):849-865

[21] Čanić S, Muha B, Bukač M. Stability of the kinematically coupled β-scheme for fluid-structure interaction problems in hemodynamics. International Journal of Numerical Analysis and Modeling. 2015;**12**(1):54-80

[22] Rybak I, Magiera J, Helmig R, Rohde C. Multirate time integration for coupled saturated/unsaturated porous medium and free flow systems. Computational Geosciences. 2015;**19**(2): 299-309

[23] Zhang L, Cai M, Mu M. A multirate approach for fluid-structure interaction computation with decoupled methods. Communications in Computational Physics. 2020;**27**(4):1014-1031

[24] Badia S, Codina R. Unified stabilized finite element formulations for the stokes and the Darcy problems. SIAM Journal on Numerical Analysis. 2009; **47**(3):1971-2000

[25] Burman E, Hansbo P. A unified stabilized method for Stokes' and Darcy's equations. Journal of Computational and Applied Mathematics. 2007;**198**(1): 35-51

[26] Cai M, Mu M. A multilevel decoupled method for a mixed Stokes/ Darcy model. Journal of Computational and Applied Mathematics. 2012;**236**(9): 2452-2465

[27] Discacciati M, Miglio E, Quarteroni A. Mathematical and numerical models for coupling surface and groundwater flows. Applied Numerical Mathematics. 2002;**43**(1):57-74

[28] Discacciati M, Quarteroni A. Convergence analysis of a subdomain iterative method for the finite element approximation of the coupling of Stokes and Darcy equations. Computing and Visualization in Science. 2004;**6**(2–3): 93-103

[29] Discacciati M, Quarteroni A, Valli A. Robin-Robin domain decomposition methods for the Stokes-Darcy coupling. SIAM Journal on Numerical Analysis. 2007;**45**(3):1246-1268

[30] Rivière B, Yotov I. Locally conservative coupling of Stokes and Darcy flows. SIAM Journal on Numerical Analysis. 2005;**42**(5):1959-1977

[31] Girault V, Raviart PA. Finite Element Methods for Navier-Stokes Equations, Theory and Algorithms, Springer Series in Computational Mathematics. Vol. Vol. 5. Berlin: Springer; 1986

[32] Elman H, Silvester D, Wathen A. Finite Elements and Fast Iterative Solvers: With Applications in Incompressible Fluid Dynamics. Oxford University Press; 2014

[33] Nield D, Bejan A. Convection in Porous Media. Vol. Vol. 3. Springer; 2006

[34] Badea L, Discacciati M, Quarteroni A. Numerical analysis of the Navier-Stokes/Darcy coupling. Numerische Mathematik. 2010;**115**(2):195-227

[35] Quarteroni A, Valli A. Domain Decomposition Methods for Partial Differential Equations. Oxford University Press; 1999

[36] Layton W, Tobiska L. A two-level method with backtracking for the Navier-Stokes equations. SIAM Journal on Numerical Analysis. 1998;**35**(5):2035-2054

[37] Brezzi F, Fortin M. Mixed and Hybrid Finite Element Methods. New York: Springer–Verlag; 1991

[38] Taylor S, Hood P. A numerical solution of the Navier-Stokes equations using the finite element technique. Computers & Fluids. 1973;**1**:73-100

[39] Cai M, Mu M, J. Xu preconditioning techniques for a mixed Stokes/Darcy model in porous media applications. Journal of Computational and Applied Mathematics. 2009;**233**(2):346-355

[40] Kay D, Loghin D, Wathen A. A preconditioner for the steady-state Navier-Stokes equations. SIAM Journal on Scientific Computing. 2002;**24**(1): 237-256

[41] Layton W, Lee H, Peterson J. Numerical solution of the stationary Navier-Stokes equations using a multilevel finite element method. SIAM Journal on Scientific Computing. 1998; **20**:1-12

[42] Dai X, Cheng X. A two-grid method based on Newton iteration for the Navier-Stokes equations. Journal of Computational and Applied Mathematics. 2008;**220**(1):566-573

[43] Cai M, Huang P, Mu M. Some multilevel decoupled algorithms for a mixed Navier-Stokes/darcy model. Advances in Computational Mathematics. 2017:1-31

[44] Adams RA. Sobolev Spaces. New York: Academic Press; 1975

[45] Layton W, Tran H, Trenchea C. Analysis of long time stability and errors of two partitioned methods for uncoupling evolutionary groundwater–surface water flows. SIAM Journal on Numerical Analysis. 2013;**51**(1):248-272

[46] Layton W, Tran H, Xiong X. Long-time stability of four methods for splitting the evolutionary Stokes-Darcy problem into Stokes and Darcy sub-problems. Journal of Computational and Applied Mathematics. 2012;**236**:3198-3217

[47] Kubacki M. Uncoupling evolutionary groundwater–surface water flows using the Crank-Nicolson Leap-Frog method. Numerical Methods for Partial Differential Equations. 2013;**29**: 1192-1216

[48] Kubacki M, Moraiti M. Analysis of a second-order, unconditionally stable, partitioned method for the evolutionary Stokes–Darcy model. International Journal of Numerical Analysis and Modeling. 2015;**12**:704-730

[49] Chen W, Gunzburger M, Sun D, Wang X. Efficient and long-time accurate second-order methods for Stokes–Darcy system. SIAM Journal on Numerical Analysis. 2013;**51**:2563-2584

[50] Causin P, Gerbeau J, Nobile F. Added-mass effect in the design of

partitioned algorithms for fluid–structure problems. Computer Methods in Applied Mechanics and Engineering. 2005;**194**(42):4506-4527

[51] Xu J, Yang K. Well-posedness and robust preconditioners for discretized fluid–structure interaction systems. Computer Methods in Applied Mechanics and Engineering. 2015;**292**: 69-91

[52] Wu Y, Cai X. A fully implicit domain decomposition based ale framework for three-dimensional fluid–structure interaction with application in blood flow computation. Journal of Computational Physics. 2014;**258**: 524-537

[53] Badia S, Nobile F, Vergara C. Fluid–structure partitioned procedures based on Robin transmission conditions. Journal of Computational Physics. 2008; **227**:7027-7051